Dynamic Planet

Pamela Elizabeth Clark

Dynamic Planet

Mercury in the Context of Its Environment

 Springer

Pamela Elizabeth Clark
Catholic University of America Physics Faculty
NASA Goddard Space Flight Center
Greenbelt, MD 20771-691
USA
pamela.clark@gsfc.nasa.gov

Library of Congress Control Number: 2006936343

ISBN-10: 0-387-48210-5   e-ISBN-10: 0-387-48214-8
ISBN-13: 978-0-387-48210-1   e-ISBN-13: 978-0-387-48214-9

Printed on acid-free paper.

9 8 7 6 5 4 3 2 1

springer.com

This book is dedicated to my colleagues, whose unceasing efforts led to a resurgence of interest in the planet Mercury and eventually to reconsideration of return missions to Mercury despite the challenges. I would particularly like to recognize those who supported, encouraged, reviewed, and/or provided material to support our efforts, particularly, Susan McKenna Lawlor, who provided a great deal of initial input for the chapters on Mercury's atmosphere and magnetosphere, as well as Steven Curtis, Rosemary Killen, Martha Leake, Faith Vilas, Ann Sprague, Barbara Giles, Clark Chapman, Joe Nuth, Jim Slavin, Bob Strom, Pontus Brandt, Norman Ness, Drew Potter, Mark Robinson, Ron Lepping, and Bill Smyth. I would also like especially to thank the staff of the NASA Goddard Space Flight Center library and the Café 10 for providing supportive environments.

PREFACE

UNDERSTANDING THE PLANET MERCURY

Thirty years have elapsed since the one and only mission to Mercury, Mariner 10, performed three flybys of the planet, capturing moderate-resolution (100 m at best) images of one hemisphere (45% of the surface) and discovering that Mercury could be the only other terrestrial planet to have a global magnetic field and core dynamo analogous to the Earth's. At the time of this writing, the MESSENGER mission to Mercury has been launched. We are still a couple of years away from the first of the next flybys of Mercury, by MESSENGER, on its way to insertion into a nearly polar, but highly elliptical, orbit, seven years from launch. In the interim, a plethora of ground-based observations has been providing information on hitherto unseen aspects of Mercury's surface and exosphere. Furthermore, Mariner 10 data have been analyzed and reanalyzed as the technology for modeling and image processing has improved, leading to important breakthroughs in our understanding of Mercury and its environment.

Thus, we are writing this book with the realization that we are in a time of transition in our understanding of the planet Mercury. Of particular interest to us in this book is the emerging picture of Mercury as a very dynamic system, with interactions between interior, surface, exosphere, and magnetosphere that have influenced and constrained the evolution of each part of the system. Previous well-written books have compellingly emphasized the results of Mariner 10 and current ground-based measurements, with very little discussion of the nature and influence of the magnetosphere. This book will present the planet in the context of its surroundings, with major emphasis on each sphere, interior, surface, exosphere, and magnetosphere, and interactions between them.

Our organizational scheme for this book is as follows: Chapter 1 will provide an introduction to the solar system, planets, and their subsystems as dynamic interconnected systems, as well as a view of Mercury in the context of the solar system. Following this, Chapter 2 will discuss missions to Mercury, including details of the only deep-space mission to reach Mercury to date, Mariner 10, and brief summaries of the next committed missions to Mercury, including NASA's MESSENGER (launched in 2004) and ESA/ISAS Bepi Colombo (launch anticipated for 2014). Chapters 3 through 6 will include reviews of our current knowledge of and planned observations for Mercury's interior, surface, exosphere, and magnetosphere, respectively. The dynamic interactions between subsystems are also considered. Results already obtained by instruments on the Mariner 10 spacecraft and by multi-disciplinary ground-based observations will be described. Current interpretation of those results will be given, along with response, in the form of anticipated capability and scientific objectives of the planned missions. The final chapter describes the future of Mercury exploration, including a profile for a mission that has the potential to complement and enhance the results obtained from MESSENGER and Bepi Colombo. The final section also contains our overall conclusions.

In this way, we hope to lay the foundation for the next major influx of information from Mercury and contribute to the planning for future spacecraft encounters.

Greenbelt, Maryland   *Pamela Elizabeth Clark*

CONTENTS

LIST OF FIGURES

LIST OF TABLES

Chapter 1

MERCURY FROM A SYSTEMS PERSPECTIVE

1.1 MERCURY IN CONTEXT

Less is known about Mercury than about any other inner planet because it is a more challenging target in so many respects. Before 1974, studies of Mercury involved astronomical observations from which its orbit, rotation period, radius and mass were determined. The Mariner 10 Mission in 1974 produced intriguing results, which will be discussed in detail in the next chapters, but provided coverage for only one hemisphere.

Mercury, the innermost member of the Solar System, is never more than 28^0 from the Sun as viewed from the Earth. Consequently, Mercury is relatively difficult to observe: the planet can only by observed with terrestrial telescopes during thirty to forty days per year. Such ground-based observations are made through the Earth's atmosphere via a long path length, either at twilight or close to dawn. At twilight, the effects of atmospheric refraction and turbulence can present a significant 'seeing' problem. Under daylight conditions, the contrast between the bright disk of the planet and the background sky is low.

1.2 PHYSICAL AND ORBITAL MEASUREMENTS

Table 1-1 presents a general summary of current physical and orbital data for Mercury in the context of other terrestrial planets (Weissman et al, 1999; Chamberlain and Hunten, 1987; Langel et al, 1980; Ness et al, 1979; Russell et al, 1974, 1979). This compilation derives from many decades of work by planetary investigators. The data are derived from both Earth-based and in-situ (Mariner 10) observations. Some of the difficulties experienced in arriving at these parameters are outlined below to indicate representative problems experienced in studying Mercury. Only solar system bodies (Mercury and Venus) describe orbits that are located between the Sun and the Earth and these two bodies are referred to as the *Inferior Planets*. Because of their locations of their orbits, the Inferior Planets display phases similar to those of the Moon.

Table 1-1. **Mercury's Planetary Characteristics in Context**

Terrestrial Planets Physical Characteristics Comparison							
planet	Total Mass (g)	Mean density (g/cm^3)	Surface gravity (cm/s^2)	Escape velocity (km/s)	Surface T extremes (K)	Normal albedo (5° phase angle)	Magnetic dipole moment (J/T)
Mercury	3.3 x 10^{26}	5.4	370	4.25	90-470	0.06	4.9 x 10^{19} (1)
Moon	3.3 x 10^{26}	3.3	162	2.38	50-123	0.07	<1.3 x 10^{15} (2)
Earth	6.0 x 10^{27}	5.5	978	11.19	260-310	0.30	7.9 x 10^{22} (3)
Mars	6.4 x 10^{26}	3.9	367	5.03	130-20	0.27	<2.1 x 10^{18} (4)
Refs: 1) Ness et al, 1979; 2) Russell et al, 1974; 3) Langel et al, 1980; 4) Russell et al, 1979.							

Terrestrial Planets Orbital Characteristics Comparison							
planet	Sidereal period (days)	Rotation period (days)	Spin:Orbit Resonance	Perihelion (AU)	Aphelion (AU)	Obliquity (degrees)	Orbital inclination (degrees)
Mercury	87.97^6	58.65	3:2	0.308	0.467	0.	7.0
Moon	27	27	1:1	0.98	1.02	1.5	5.1 (Earth)
Earth	365	1.0	365:1	0.98	1.02	23	0.0
Mars	686	1.0	686:1	1.38	1.67	25	1.9

1.3   DIFFICULTIES AND ANOMALIES UNCOVERED IN OBSERVING MERCURY

Attempts to determine Mercury's basic physical properties led to the determination of its unexpectedly large mass and implied high density. The mass of Mercury was first derived by a German astronomer Johann Franz Encke in 1841 when he measured the perturbations produced by Mercury on a comet that has since been given his name. The measurement remained controversial until the mass and size of this body were later more accurately determined from combined ground-based and Mariner 10 observations (Lyttleton, 1980, 1981; Branham, 1994; Anderson et al, 1987). A major implication for this measurement was that the density of the planet was much higher than models could explain, implying disproportionately greater iron abundance and core size. This will be discussed in full in Chapter 3 on Mercury's Interior.

Methods of computerized location and tracking have recently increased the probability of successfully observing Mercury from the Earth under daylight conditions. As a direct result of not only these technological advances but of the unanticipated scientific advances made during the Mariner 10 mission in 1974 (NASA Atlas of Mercury), a more vigorous and multi-faceted ground observation program for Mercury has ensued in the last three decades. **Table 1-2** is a summary of findings on the figure, orbital, and surface properties of Mercury made from ground-based measurements.

Table 1-2. **Ground-based Observations Contributions**

Type of Observation/Target	Finding
Visible/Near to Mid IR (Chapters 3, 4) Regolith Compositional Properties	Minimal (1-2%) iron in iron-bearing silicate minerals in regolith
Visible Spectrometer Lines, Images (Chapter 5), Exosphere	Spatial and Temporal characterization of Na, K, Ca exosphere
Thermal, Microwave, Radar Figure and Orbital Properties (Chapter 1)	Radar ranging for figure and dynamic properties, confirm General Relativity, radar imaging and topography confirm impact, tectonic, polar features
Regolith Physical Properties, Surface topography, morphology (Chapter 4)	Thermal, Microwave reflectivity, polarization ratios indicates lunarlike regolith, polar volatiles

In 1965, on the basis of Doppler radar measurements made at Arecibo Observatory, Colombo and Shapiro (1996) unexpectedly demonstrated that Mercury exhibits a rotation period of 58.6 days, a value that is exactly two thirds of its 88 days orbital period. Yet, previous optical observations apparently indicated that the rotation and orbital periods were the same. Why the apparent difference between the result obtained using radar and that determined optically? The inference was drawn, from the radar based result, that Mercury rotates three times about its axis during every two of its orbits about the Sun, and that after three synodic periods, the same face of Mercury viewed at the same phase, will be presented to Earth based observers. Three synodic periods is also the time interval between the most favorable conditions for viewing Mercury telescopically from the northern hemisphere. An astronomer working in this hemisphere at favorable times would consistently track 88-day passages of characteristic surface markings for up to six consecutive years. An obvious drift in the positions of the markings would thereafter, for geometrical reasons, become apparent. However, because optical maps of Mercury's surface features were typically made at particular observatories within the compass of programs that extended over just a few years, a set of drawings of surface features made at a particular site could, against this background, convincingly, but erroneously, suggest an 88-day rotation period for Mercury.

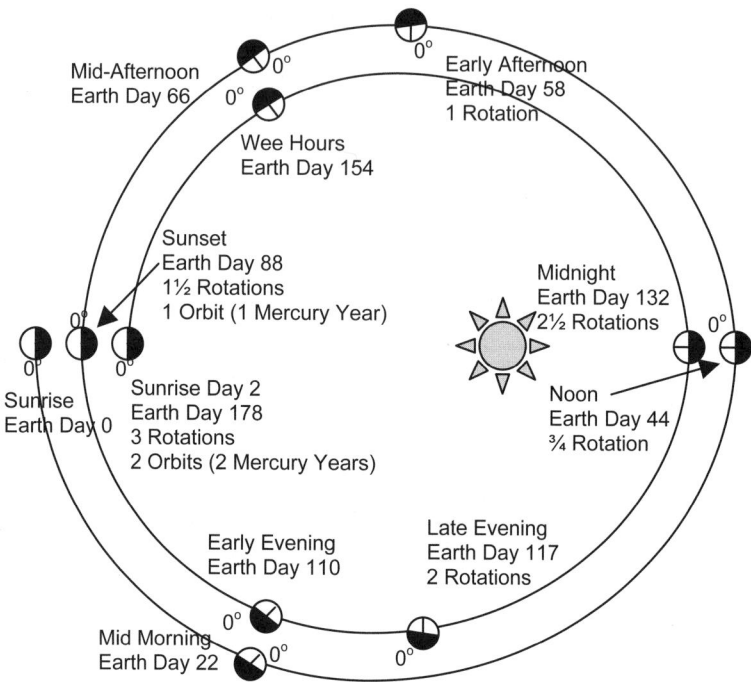

Figure 1-1. Mercury spin-orbit coupling: Mercury's 3:2 spin-orbit coupling is illustrated by this diagram of the planet's illumination as a function of position in its orbit. In the starting position at aphelion, it is sunrise at one of the terminators. After half an orbit, it is noon at that terminator. After a complete orbit, Mercury has rotated 1.5 times, and it's sunset at the same terminator. After two complete orbits, Mercury has rotated three times and it is sunrise again at the same terminator. (After Strom, 1987. Courtesy of Robert Strom.)

Mercury has the most eccentric and inclined orbit of all the terrestrial planets, and, as a result, displays the greatest variation in its heliocentric distance. Spin orbit coupling combined with the pronounced orbital eccentricity of Mercury result in the planet first presenting one particular hemisphere, then the one opposite to it, to the Sun during successive perihelion passages **(Figure 1-1)**. Mercury's prime meridian was chosen to pass through the sub-solar point at the first perihelion passage that occurred in 1950; thus, central meridians 0^0 and 180^0 always face the Sun at perihelion.

With an obliquity close to zero degrees, Hermean latitudes experience diurnal, but not seasonal, changes in temperature. Overall, Mercury's proximity to the Sun, lack of an insulating atmosphere and long day cause it

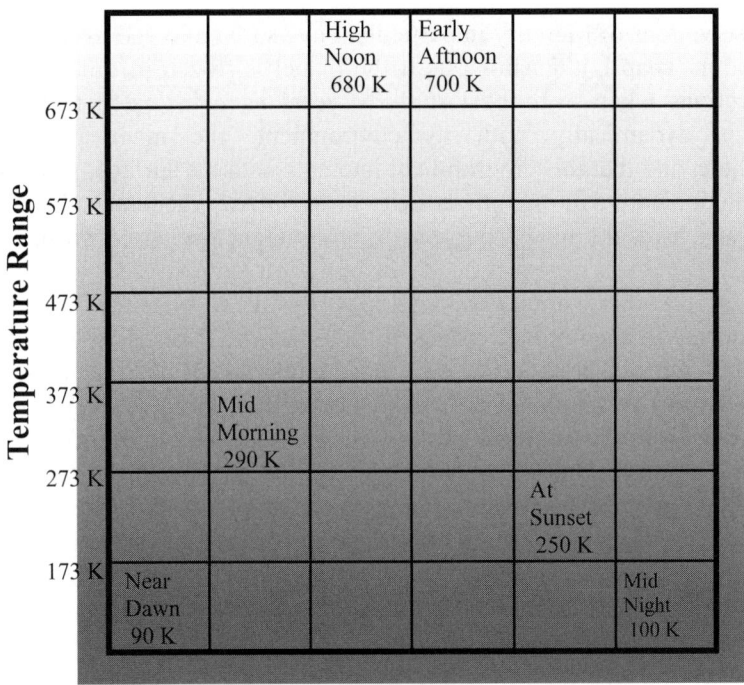

Time of Day

Figure 1-2. Mercury's extreme temperature cycle.

to display the greatest difference among the planets of the Solar System between its day and night temperatures. Diurnal Temperature cycles as a function of a Mercury day (one spin) (Strom and Sprague, 2003) are shown in **Figure 1-2**.

When the planet is closest to the Sun, Mercury's surface temperature reaches about 740 K. The 0^0 and 180^0 longitude meridians are 'hot poles' as a result of the enhanced heating received at these locations as the planet passes through the perihelion of its eccentric orbit. A signature of the temperature enhancement concerned is readily observed from the Earth in microwave radiation. .The central meridians 90^0 and 270^0 face the Sun only at aphelion and the maximum temperature thereby reached is significantly lower (about 525^0 K). On Mercury's night side, the Sun does not shine for months at a time due to the slow rotation of the planet. The surface temperature can thus drop to about 90 K.

1.4   A PLANET AS A SYSTEM OF SUBSYSTEMS

Observations of Mercury in particular have so far allowed us to capture its essence in 'snapshot' fashion, as if the planet is frozen in time. However, observations taken over the last thirty years have indicated that Mercury interacts dynamically with its environment, the magnetosphere and exosphere, and that this environment interacts with the surface. Most likely a dynamo, indicative of ongoing activity in the interior, generates the magnetic field, which results in the presence of the magnetosphere. To understand this active planet, we must see it in terms of temporal as well as spatial variations, in other words, as an interactive system of systems.

A planet is a complex system of subsystems. By using a systematic approach, we can bring order out of the chaos of fascinating idiosyncratic details a particular planet displays (Lewis and Prinn, 1984). This is a challenge. Natural environments are complex. When we observe them, we are not looking at controlled laboratory experiments, where we can carefully vary one parameter at a time, but at many known and unknown simultaneously varying parameters. We must hone in on and capture the most essential ones in order to constrain variation over orders of magnitude. We must be around long enough to observe events that we can't control and that perturb the system, and deal with incomplete or noisy observations in the process. We must be willing not only to abandon old paradigms and formulate new ones, but to consider 'multiple working hypotheses', a very unsettling challenge for most human beings who love certainty.

Energy is transferred between subsystems and the 'work' of the system is done at interfaces where changes of state or phase occur. As described in **Table 1-3**, states include solids, liquids, gases, and, the most ubiquitous state in the universe, plasma. The subsystems include interior, surface, exosphere/atmosphere, and magnetosphere, the origin and generic properties of which are described in **Table 1-4** and in more detail further on.

1.5   TYPES OF SYSTEMS

Systems can be in equilibrium or not in equilibrium. Non-equilibrium models, accounting for most natural systems, range from steady state models where conditions are apparently at equilibrium locally to dynamic models, which exhibit predictable or chaotic variations (Lewis and Prinn, 1984) (**Table 1-5**).

Equilibrium systems are ultimately balanced, in an apparently unchanging non-evolutionary state with no spatial gradients. Nor are there temporal changes or fluxes. Many natural systems that are not actually in this state are treated as equilibrium systems if temporal changes are slow by human standards. State functions are constant, entropy is maximum, and free energy is minimum.

Steady state systems exhibit at least minimal temporal and spatial changes, often at a small enough scale so that, locally, they can be treated as equilibrium systems. However, they do exhibit flow and gradients and non-equilibrium thermodynamics overall. Entropy is no longer maximum and free energy no longer minimum. Temperature variation in the solar system varies across a gradient as a function of distance from the sun. Apparent equilibrium at the top of each planet's atmosphere is governed by its surrounding external state in terms of temperature.

Table 1-3. **States of Matter**

State	Solid	Liquid	Gas	Plasma
Example	Ice H_2O	Water H_2O	Steam H_2O	Ionized Gas $H_2 > H+ + H++2e-$
Description	Molecules Fixed in Crystal Lattice	Molecules not fixed, free to move in confined space	Molecules Free to Move in Large Space	Components (ions and electrons) move independently in large space
Vary Temperature	Coldest ---→ Hottest			
Vary Pressure	Highest ---→ Lowest			

Table 1-4. **General Description of Major Planetary Subsystems**

Subsystem, Structure, Phase Transitions	Sources, Sinks, Processes
Interior Core/Mantle/Crust Boundaries at phase changes Typically liquid to solid	Mixing between layers, Infall meteoritic material, volcanotectonic activity through melting, creep, plastic and elastic deformation
Surface Solid rubble (regolith) Interface between Crust and Exosphere	Interior (volcanic eruption and tectonic displacement) and Exterior (meteoritic, comet, particle) infall, interior-driven resurfacing and cratering (removal and deposition, gardening and space weathering) processes
Exosphere/Atmosphere Interface between surface (neutrals) and plasma (ionized forming ionosphere) layers Typically gas and plasma	Surface molecules, atoms, and ions produced by interaction with plasma, lost through escape and adsorption
Magnetosphere: the somewhat leaky boundary between interplanetary and planetary magnetic field plasmas Magnetopause is Interface with InterPlanetary Field and Magnetotail is Wake	Particles transferred across magnetic field interfaces, in and out of subsystem, by reconnection and across atmosphere and surface boundaries by charge and momentum transfer

Dynamic systems can be cyclical, evolutionary, or catastrophic (Lewis and Prinn, 1984). Cyclical temperature variation cycles, which occur simultaneously, include the diurnal and annual cycles related to a planet's spin and revolution, respectively. Evolutionary systems change slowly due to long-term systematic variations in external conditions superimposed on a steady state. Both cyclical and evolutionary systems are uniformitarian in nature, exhibiting regular, incremental variation in a predictable direction. On the other hand, catastrophic systems exhibit rapid, large-scale, unanticipated changes resulting from major transitions in surrounding environment conditions.

Table 1-5. **System Model Characteristics**

System Type	Thermodynamics	Dynamics	Characteristics
Uniformitarian	Equilibrium	Constant	No spatial or temporal gradients
	Non-equilibrium	Steady State	Locally, approaches equilibrium, Overall incremental temporal and spatial gradients
		Dynamic	Cyclical, repeated patterns in changes induced by changed in external conditions
			Evolutionary, predictable trends in changes induced by gradual changes in external conditions
Catastrophic			Catastrophic, large-scale, unpredictable, unrepeatable changes induced by major shifts in external environment

1.6   IN THE BEGINNING: SOLAR NEBULA SYSTEM FOR PLANET FORMATION

The solar nebula represents a spatial and temporal model for planet formation. The system is powered by nuclear energy. Converted to its gravitational, kinetic, and rotational forms, this energy acts as the 'mechanical resource' that can be harnessed to do the work of planetary formation (Elder, 1987). Energy is transferred across interfaces between matter in changed state or phase: e.g., plasma to gas, or high pressure to lower pressure forms of a mineral. Plasma is the most ubiquitous form of matter in the relatively hot and low pressure universe, but, locally in the nebula, conditions get cool and dense enough for matter to coalesce to form solids.

What are the initial conditions for planet formation? (e.g., Lewis and Prinn, 1984; Lewis, 2004; Encrenaz et al, 2003) Recent observations of extra-solar systems in formation have helped to refine the models and constrain the assumptions of these conditions. Planetary systems begin as denser than average gas clouds in hydrogen rich arms of relatively young galaxies (Elder, 1987). Bulk composition, including volatile content and oxidation state, the availability of energy sources (achieving sufficient heat), and the mass of the starting material (achieving sufficient density) are parameters of primary importance in influencing the stability, condensation, and vaporization of small grains in the early nebular phase, as well as the later differentiation and crystallization of larger bodies in the planetary formation stage. Both processes are essential to planet formation.

The planetary formation model initially developed by Safronov (1972) is currently the prevailing paradigm, largely because it best explains the current observations of solar systems outside of our own. In addition, it is the planetary origin model which best accounts for the following observations:

(1) the observed flattening of the nebular disk during planet formation,

(2) distribution and compositional variations between planets including the observed loss of volatiles in the inner solar system and solar abundances among the oldest measured material,

(3) order and duration of events as derived from radiochemical dating measurements (including 10^8 year planet formation period after initial disk formation),

(4) the observed deuterium loss in the sun, where it was used up after nebular disk formation and accretion,

5) the transfer of angular momentum from the sun to planets which, according to recent models, required only small drag and mass loss by the rotating sun,

6) observed small body dynamics including asteroid belt formation.

In Safranov's model (1972), as illustrated in **Figure 1-3**, gas condensed into a diffuse, homogeneous dust cloud which collapsed and accreted heterogeneously relative to the nebular center in response to temperature and pressure gradients. Although temperature varied rapidly with distance, dust, acting as a good insulator, preserved temperature gradients within the accretion disk, and convection transferred heat under steady state conditions. Dust settled toward the midplane with resulting mass distribution inhomogeneities and gravitational instabilities.

Large-scale fluctuations in the output of the early sun brought on the next phase of nebular development (Lewis and Prinn, 1984; Lewis, 2004; Encrenaz et al, 2003). Such events resulted in sudden short energetic pulse or pulses, causing very rapid remelting and solidification of the dust throughout the solar nebula. This event generated the igneous spheroid droplets (chondrules) observed in chondritic meteorites. Chondrules accumulated still-condensing dust layers on their surfaces, and began to coalesce into planetesimals. Subject to drag with the surrounding gas, planetesimals slowed down and spiraled in toward the sun until they grew to a size of about ~1 km in diameter where, due to less surface area per volume, the drag became negligible. There, these coalesced bodies stayed and eventually grew into planetoids, bodies tens of kilometers or greater in diameter.

Stages of Solar System Formation

Irregularities in Plasma Distribution: :
Leading to Nebula Formation

Plasma -> Gas

Irregularities in Nebula
Leading to Nebular Disk Formation

Plasma + Gas + Liquids

Irregularities in Nebular Disk:
Leading to Coagulation into Planetesimals

Plasma + Gas + Liquids + Solids

Irregularities in Planetesimal Distribution:
Leading to Planetoid Formation and Growth

Solids, liquids, and gases differentiate on basis
of temperature and pressure.

Figure 1-3. Stages of solar system formation, showing systematic changes in distribution and states of matter induced by heterogeneous distribution of matter as discussed in text.

As accretion continued during the planetoid formation stage, compositional boundaries blurred, grains became well-mixed, forming polymict breccias with chondrules, known as chondrites. The sun experienced a sudden, high temperature explosive event known as the T Tauri phase which caused gas to be swept out dragging dust and the smaller bodies along with it.

The larger bodies remained, becoming today's observed planets, but lost their original surrounding gas envelopes, primordial atmospheres, in the process. Closest to the center of the nebula, more severely heated bodies became the differentiated achondrites in which the chondrites had remelted and disappeared. Many bodies remained as chondrites, modified to a degree determined by their distance from the nebular center. On this side of the asteroid belt, chondrites have been subject to some increase of temperature and pressure, metamorphosed to some degree, are coarse grained and low in volatiles. Elsewhere, chondrites remained 'primitive', fine grained, volatile rich, and with low temperature mineral assemblages. Oxygen and other light element isotope ratios are diagnostic for major meteorite classes. Discrete variations within meteorite classes has led to the concept that several larger parent bodies were the source for each meteorite class.

Temperature at which solid dust grains form depends to a first approximation on composition, as indicated in **Table 1-6** (e.g., Lewis and Prinn, 1984; Encrenaz et al, 2003; Lewis, 2004), with secondary variations in this order depending on the oxidation state. Refractory oxides solidify at the highest temperatures, followed by metallic FeNi, then Mg-rich silicates and quartz, followed by alkali and aluminum bearing silicates. Reactions with iron form iron sulfides and oxides occurs at somewhat lower temperatures. Metamorphic mineral indicators such as tremolite and serpentine form at lower temperatures, followed by formation of volatile ices. Dust formation is accompanied by volatile entrapment. Radiogenic helium is retained at the highest temperature, rare gases at a somewhat lower temperature, and simple molecules at much lower temperatures.

Table 1-6. **Accretion and Volatile Retention as a Function of Temperature (Lewis, 2004)**

Temperature	Solidification/Volatile Retention
>1600K	Refractory (REE) Oxides
<1600K	Radiogenic Helium
>1400K	Metallic FeNi from Fe and Ni vapor
<1400K	Rare Gases
>1250K	Mg-rich olivine and pyroxene, quartz by reaction SiO, Mg, water
>1000K	Alkali Aluminum silicates by reaction alkali vapor with high temperature Calcium Aluminum silicates
<1000K	Argon40
>600K	Corrosion Fe by reaction with H_2S to form FeS Oxidation Fe to form FeO-bearing silicates
>350K	Tremolite (calcic amphibole) from enstatite, diopside, water vapor Serpentine from water, ferromagnesian minerals
>200K	Ices

The elegantly simple and compelling idea that formation and composition of solid grains depend on temperature, and thus on solar distance, is the basis of the equilibrium condensation model (Lewis, 1988). An underlying assumption of the equilibrium condensation model is that the accretion process started at the highest temperature, with all matter in vapor form, and then slow cooling occurred. Of course, an apparent relationship between composition and solar distance has been observed for planets of known composition. This assumption is reinforced by the fact that meteorite classes have distinctive oxygen isotope ratios, which are temperature dependent. Observations which apparently depart from this model in its purest form, which include those of the planet Mercury, have led to the modification of the initial model as well as the development of different models for planet formation, which will be discussed in detail in Chapter 3.

Stages of Planet Formation

| Opportunistic Grain Aggregation | Partial Melting, Fractional Crystallization into Magma Ocean | Differentiated Body Cooling from partially molten into solid |

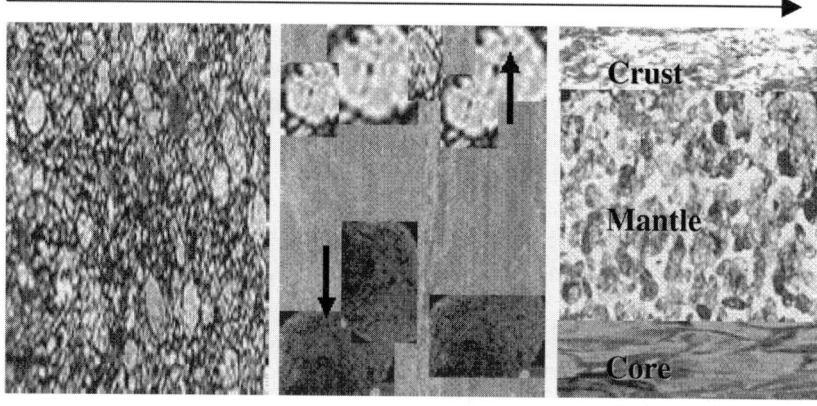

Figure 1-4. Stages of planet formation resulting from initial accumulation of grains followed by partial melting resulting in fractional crystallization and growing heterogeneity to produce a differentiated body, as discussed in text.

1.7 INTERIOR AND SURFACE FORMATION: SOURCES, SINKS, PROCESSES

Planets themselves are complex systems as they develop an interior structure which results from differential flow and fractional crystallization of materials in a range of solid, liquid, and gaseous states and phases (Elder, 1987), as illustrated in **Figure 1-4**. These state transitions and their resulting structures depend on composition, temperature, and pressure conditions as a function of depth. As cooling occurs, prevailing states and interior structure change. Gases solidify and shrink around dusty liquid cores. Solid crust grows and liquid interior shrinks as crust, mantle, and core segregate until eventually, the body is completely solid. Radiogenic materials are a primary heat source. Heat transfer through direct conduction is inefficient. The transfer of heat through the interior occurs primarily through magma (liquid) convection and solid (heated fluids within mantle) convection. The Nusselt Number (Actual Heat Transfer/Just Conductive Heat Transfer) is an indication of how much geological or radiogenic source activity is present.

What are the heat sources available for differentiation? Long-lived radionuclides would be a primary source. Volatiles could also play a role by lowering the melting temperature of the fluid containing them. While being

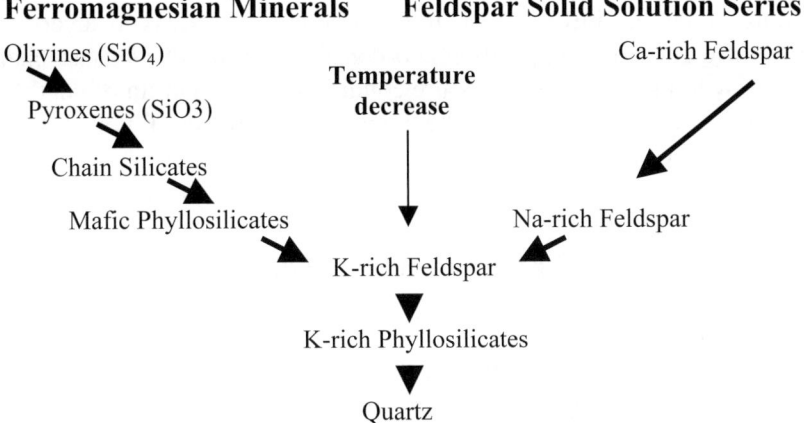

Figure 1-5. Bowen's reaction series. (N.L. Bowen, 1907–1980.)

outgassed, volatiles raise the temperature of the surrounding curst while not completely melting it. The release of gravitational and chemical potential energy in the interior would also be a source of heating. In addition, ohmic heating (resistance) resulting from solar wind charging and electric current flow at the surface could be significant for bodies that were either close to the sun or poor conductors or both.

How do heating and cooling processes result in differentiation of the interior into core, mantle, and crust? What implications are there of this process from crustal composition, which we can observe or sample directly? Many workers believe that initially (Encrenaz et al, 2003), the planetoid is a fluid at high temperature, in other words, a 'magma ocean' (Wood, 1972). As the temperature falls, crust formation begins. Solid 'bergs' made of the highest melting temperature mineral assemblages solidify. The relatively low density ones float on the surface, while high density ones sink to form mantle. In the middle is the remaining undifferentiated liquid, progressively changing in composition as the temperature falls and fractional crystallization occurs. Composition of solidifying material also changes in a predictable way as the composition of its remaining source material changes. The order in which such assemblages form as temperature falls and crystallization occurs can be approximated by the Bowen reaction series (N.L. Bowen, 1907-1980) shown in **Figure 1-5**. In truth, such a 'magma ocean' process does not produce a completely uniform, homogenous crust. Slabs rise, fall, and tilt. Horizontal and vertical variations in composition occur when density differences result from physical differences in texture from variation in eruption styles.

In the interior, below the crust, materials behave like viscous fluid, creeping under the influence of gravity. Boundaries between layers occur due to differences in composition and density (as a result of phase changes induced by temperature and pressure conditions as a function of depth). An incipient atmosphere forms as the interior is emptied of volatiles which arrive hot at the surface.

1.8   ATMOSPHERE FORMATION: SOURCES, SINKS, AND PROCESSES

The atmospheres associated with solar system bodies today are not the primordial ones enveloping coalescing grains (Lewis and Prinn, 1984). Such atmospheres were lost during the T Tauri event which occurred after the early planets formed. What we see today are atmospheres that have resulted through various acquisition and loss processes.   Volatiles have been outgassed through the heating, melting, and differentiating of the interior described above. When the net loss of these volatiles results, exospheres, like Mercury's, can form from escaping ionizing volatiles. This fact was clearly illustrated by the inability of Mariner 10 to find the anticipated (primordial) atmosphere around Mercury (Kumar, 1976). H and He abundances were low enough to have had a strictly solar origin.   The minimal O could have resulted from sputtering. Na and K found more recently could also result from sputtering. (See Chapter 5 for details.)

Table 1-7. **Some Primordial Gas Equilibria Driven Right by H Loss**

$CH_4 + H_2O \rightarrow CO + 3H_2$
$CO + H_2O \rightarrow CO_2 + H_2$
$CH_4 \rightarrow C \text{ (graphite)} + 2H_2$
$2NH_3 \rightarrow N_2 + 3H_2$
$H_2S + 2H_2O \rightarrow SO_2 + 3H_2$
$8H_2S \rightarrow S_8 \text{ (solid)} + 8H_2$

Table 1-8. **Comparison of Current Terrestrial Planetary Atmospheres**

Terrestrial Planet	Major Constituents	Pressure at Surface
Mercury Exosphere	H, He, O_2, Na, K, Ar	Trace
Venus Atmosphere	Major CO_2, N2 Minor/Trace SO_2, Ar, H_2O, CO, He, Rare Gases	92 Bars
Earth Atmosphere	Major N_2, O_2, Ar Minor/Trace CO_2, H_2O, SO_2, CO, CH_4	1 Bar
Mars Atmosphere	Major CO_2, N_2, Ar, O_2 Minor/Trace Co, H_2O, Rare Gases	0.07 Bar

The loss of primordial H and He drives equilibrium **(Table 1-7)** (Encrenaz et al, 2003)in the direction of producing oxidized gases including CO_2, now found in the atmospheres of terrestrial planets that can retain their volatiles. Excess C, due to its greater solar abundance, is deposited as graphite. O is used up in producing H_2O.

Whether volatiles released from the interior can form a steady state atmosphere depends on the dynamic balancing of sources and sinks. As illustrated in **Table 1-8**, three out of four terrestrial planets have managed to create a 'steady state' atmosphere. Such a result can only occur if production is fast relative to transport, diffusion, and loss via escape, destruction, or dissociation by high energy particles or UV radiation to produce unstable ions, free radicals, and electrons. This topic will be discussed in much more detail in Chapter 5. Meanwhile, the interaction of such particles at the surface, via dissolution, photocondensation, or other reactions, results in 'space weathering' of the surface, described in more detail in Chapter 4.

1.9   MAGNETOSPHERE FORMATION: SOURCES, SINKS, AND PROCESSES

The interplanetary medium is not empty but filled with ionized particles, or plasma, and dust (Encrenaz et al, 2003). Most of the particles are from the sun, and they interact with surrounding solar system objects. Such dynamic interactions are observed in extra-solar systems as well.

This interplanetary plasma, known as the solar wind, originates in the solar corona, escaping from dark areas known as coronal holes where magnetic field lines open. This plasma interacts with planets in a manner illustrated in **Figure 1-6** (Encrenaz et al, 2003). Bodies lacking both magnetic fields and atmospheres, like the Moon, are not good conductors. The body acts as an insulator. The solar wind doesn't encounter magnetic lines of force and plasma doesn't accumulate on the upstream side, but instead ions and electrons progressively diffuse into the empty cavity downstream, creating differential charging on the day and night sides and fields that can loft dust near the terminators (Stubbs et al, 2006). In the case where an atmosphere is encountered, the outer layer of the atmosphere is ionized by the solar wind to form an ionosphere which acts as a spherical conductor. If the body has an atmosphere but lacks an intrinsic magnetic field, the interplanetary magnetic field lines are perturbed around the object to form a pseudo-magnetosphere with an ionopause rather than a magnetopause at the point of encounter. In bodies with intrinsic magnetic fields, magnetospheres form and are maintained. Charged particles flow around the body, through the ionosphere if the body has an atmosphere, or along the surface if it does not. Within ionospheres, magnetic flux tubs form

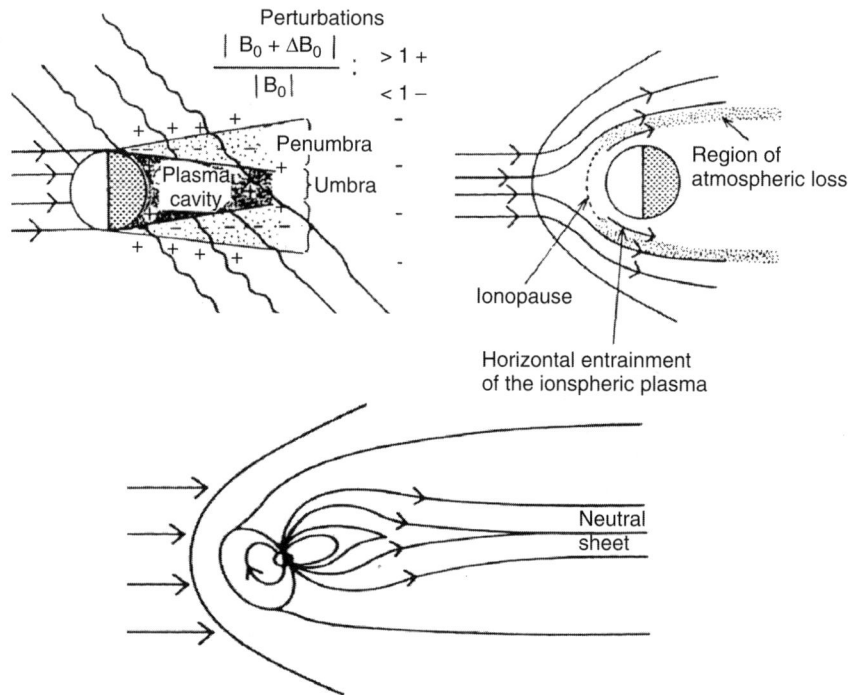

Figure 1-6. Interaction between planetary bodies and the solar wind for, clockwise from top left, bodies without atmospheres or magnetic fields, such as the Moon; bodies with atmospheres but no magnetospheres, such as Venus; and bodies with magnetospheres. (Based on work of R.P. Lepping. Reprinted with permission of Springer-Verlag.)

along closed field lines, enclosing an area known as the plasma sphere through which charged particles can diffuse. Particles move along field lines, get diffused and connect the ionosphere to the internal dynamos at the poles and then, acting as a polar wind, move by magnetic convection into the magnetotail, reconnecting to form ring currents and aurorae on the nightside and then to be recycled into the plasmasphere. Direct solar wind interaction with the surface can occur in planets with small or non-existent magnetic fields and atmospheres.

A magnetosphere forms an envelope around the planet, inside of which particles are subject to dynamic control by the planet's intrinsic magnetic field. A Bow shock is created where the plasma first comes into contact with the Earth's magnetic field. Just inside the Bow Shock, the magnetosheath heats, slows, and deflects the solar wind, but some of it leaks into the envelope, through the magnetopause. The dynamic behavior of these particles is then controlled by the magnetosphere. This leaky behavior is the result of partial reconnection, the change in orientation of electrons accompanied by shock when they are transferred from one system to another. The wake which forms on the side opposite the sun is known as the magnetotail.

1.10   SUMMARY

First astronomical observations over the last century and then Mariner 10 mission in 1974 provided a snapshot of Mercury showing spatial relationships. The picture is far from complete for a body as dynamic as Mercury. Mercury apparently has an active dynamo implying some level of geothermal activity. The planet interacts dynamically with its rapidly changing surrounding environment. The exosphere and magnetosphere processes are dynamic with the interior, surface, solar wind and exosphere generated plasmas acting as modifying agents, sources, and sinks. All these conditions imply temporal changes and have led us to use a system of systems approach to studying Mercury. Here we discuss the planetary nebula as a dynamic system for planet formation. We also introduce the generic concept of the interaction between spheres, or subsystems which connect at interfaces where phase changes occur: within the interior between core, mantle, and crust, between crustal surface and gaseous exosphere, between exosphere and plasma-driven magnetosphere. Each subsystem is treated in terms of sources, sinks, and processes. This chapter acts as an introduction to the detailed discussion of such interactions on Mercury, with chapters on each of its subsystems.

1.11   REFERENCES

Anderson, J. D., G. Colombo, P. B. Esposito, E. L. Lau, and G. B. Trager, The mass, gravity field, and ephemeris of Mercury, *Icarus*, **71,** 337-349, 1987.

Bowen, N.L., Papers1907-1980, Compiled by J. Snyder, Geophysical Lab, Carnegie Institute, Washington DC, 2004.

Branham, The mass of Mercury, *Planet. Space Sci,* **42,** 213-219, 1994.

Chamberlain, J.W. and D.M. Hunten, *Theory of Planetary Atmospheres*, 2nd Ed. (Publ. Academic Press), 481 p., 1987.

Colombo, G. and I. I. Shapiro. The rotation of Mercury, *Astrophys. Jour.*, **145**, 296-307,1966.

Elder, J., *The Structure of the Planets* (Publ. Academic Press), 210p., 1987.

Encrenaz, T., J.P. Bibring, M. Blanc, M.A. Barucci, F. Roques, P. Zarka, *The Solar System*, (Publ. Springer), 512 p., 2003.

Langel, R., R. Estes, G. Mead, E. Fabiano, and E. Lancaster. 1980. Initial geomagnetic field model from magsat vector data. *Geophysical Research Letters*, 7(10):793-796, 1980.

Lewis, J.S., Metal/silicate fractionation in the solar system *Earth Plan Sci Lett,* **15,** 286- 292, 1972.

Lewis, J. S., Origin and composition of Mercury. In *Mercury,* Eds. Vilas, Chapman, and Matthews (Publ. Univ. Arizona Press)*,* pp. 651-666, 1988.

Lewis, J.S., *Physics and Chemistry of the Solar System,* 2nd Ed. (Publ. Elsevier), 655p., 2004.

Lewis, J.S. and R.G. Prinn, *Planets and their Atmospheres Origin and Evolution* (Publ. Academic Press), 470p., 1984.

Lyttleton, R.A., History of the mass of Mercury, *QJR Astro Soc*, **21**, 400-413, 1980.

Lyttleton, R.A., More thoughts about Mercury, *QJR Astro Soc*, **22**, 322-323, 1981.

NASA, Atlas of Mercury, SP423, 1975.

Ness, N. F. The magnetic field of Mercury, *Physics of the Earth and Planetary Interiors*, 20:209-217, 1979.

Russell, C., P. Coleman, and G. Schubert, Lunar magnetic field: permanent and induced dipole moments. *Science*, 186:825-826, 1974.

Russell, C. T. The Martian magnetic field. *Physics of the Earth and Planetary Interiors*, 20:237-246, 1979.

Safranov, V.S., Accumulation of the Planets. in *Origin of the Solar System*, Ed H. Reeves (Publ. CNRS Paris), 89-113, 1972.

Strom, R., *Mercury: The Elusive Planet*, Smithsonian Institution Press, Washington DC, 1987.

Strom, R. and A. Sprague, Chapter 3 in *Exploring Mercury: The Iron Planet*, Springer-Verlag, 2003.

Stubbs, T.J., R.R. Vondrak, W.M. Farrell, A dynamic fountain model for lunar dust, *Advances in Space Research* 37, 1, 59-66, 2006.

Weidenschilling S.J., Iron/silicate fractionation and the origin of Mercury, *Icarus*, **35**, 99-111, 1978.

Wetherill, G., Accumulation of Mercury from planetesimals. In *Mercury*, Eds. F. Vilas, Chapman, and Matthews (U. Arizona Press) 670-691, 1988.

Weissman, P.R., L.A. McFadden, T.V. Johnson, Physical and Orbital Properties of the Sun and Planets, in *Encyclopedia of the Solar System* (Publ. Academic Press), p. 961, 1999.

Wood, J.A., Thermal History and Early Magmatism of the Moon, *Icarus*, 16, 2, 229, 1972.

1.12 SOME QUESTIONS FOR DISCUSSION

1. What has been the influence of observations of extra solar systems on prevailing paradigms for planetary formation?

2. Describe a 'typical' planet in terms of the spatial and temporal features of energy and matter transfer processes.

3. Compare and contrast the utility or weaknesses of the following types of system models: Equilibrium, Steady State, Evolutionary, Catastrophic.

4. Explain systematic differences in any property among planets of our solar system.

Chapter 2

PAST AND PLANNED MISSIONS TO MERCURY

2.1 NASA'S SUCCESSFUL MARINER 10 MISSION TO MERCURY

Although acquiring ground-based astronomical observations of Mercury is difficult, visiting the planet via spacecraft to acquire observations in-situ is even more challenging. Its close proximity to the sun creates high thermal radiation and high gravity environments.

At this time, only one space mission to Mercury, NASA's Mariner 10 (Clark, 2004) has actually rendezvoused with the planet. Three encounters by Mariner 10 (M10) in 1974 and 1975 provided the first in-situ observations of one hemisphere. An especially important discovery was that Mercury has an intrinsic magnetic field, implying that the planet has a partially molten, iron-rich core and, thus, a history of extensive geochemical differentiation. However, lack of global coverage (only 45% of the surface was imaged), and the limited nature of many onboard measurements, has lead to largely unconstrained theories of Mercury's origin and history.

Table 2-1. **Mariner10 Details**

Launch	Flight

Mission Management: JPL
Launch: November 3, '73: 5:45 UTC
Launch Site: Cape Canaveral, USA
Launch Vehicle: Atlas Centaur 34
Spacecraft Mass: 503 kg
	Arrivals:
	  Venus: February 2, '74 (5768 km)
	  Mercury 1: March 29, '74 (703 km)
	  Mercury 2: September 21, '74 (48069 km)
	  Mercury 3: March 16, '75 (327 km)
	End of Mission: March 24, '75

The Mariner 10 Mission **(Figure 2-1, Table 2-1)** was launched on November 3, 1973, the first day of its scheduled launch period. The

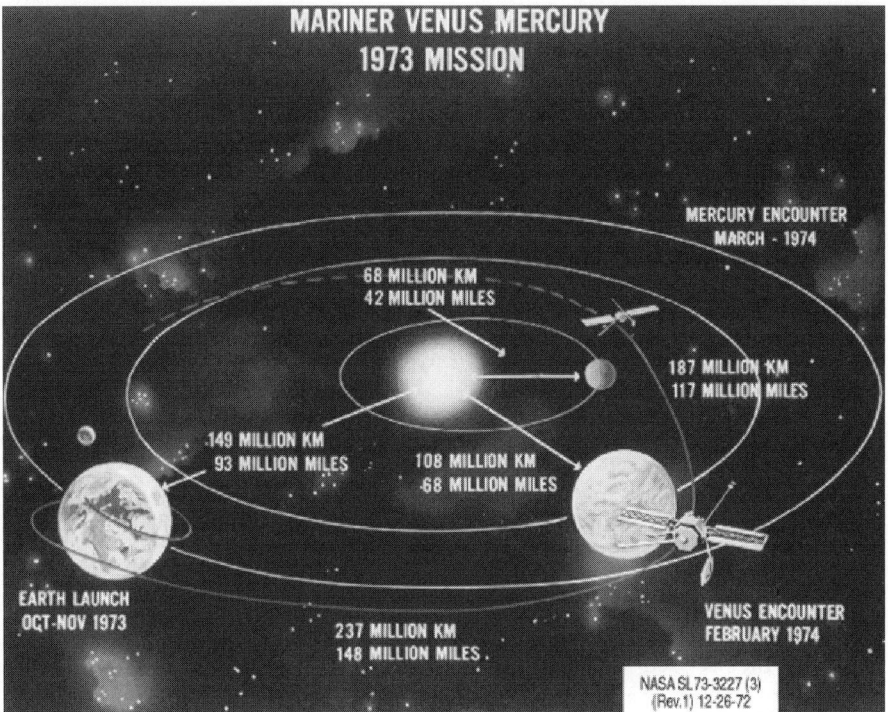

Figure 2-1. Mariner 10 mission scenario, showing the 'firsts' that are necessarily associated with every mission to Mercury: Here, these include the first encounters with the planet Mercury and the first use of the gravitational assist technique. (Found at http://www.hrw.com/science/si-science/physical/astronomy/ss/mercury/img/marinertraject.gif.)

spacecraft encountered Venus in early 1974, when it provided the first close-range measurements of this planet while also executing a gravity-assist maneuver that enabled it to later reach Mercury. Historically, Mariner 10 was the first mission to utilize a gravitational-assist trajectory, as well as the first to visit, at close range, more than one planetary target. The spacecraft was then transferred into a retrograde orbit around the sun. In this orbit, the spacecraft encountered Mercury three times. **Tables 2-1 and 2-2** list the mission firsts and details.

Table 2-2. **Mariner 10 Payload**

Instrument	PI, PI Institute
TVTab System	B. Murray, Cal Tech
IR Radiometer	C. Chase, Santa Barbara Research
UV Airglow and Occultation Spectrometers	A. Broadfoot, Kitt Peak
Radio Science and Celestial Mechanics Package	H. Howard, Stanford University
Magnetometer	N. Ness, Goddard Space Flight Center
Charged Particle Telescope	J. Simpson, U. Chicago
Plasma Analyzer	H. Bridge, MIT

The first flyby (variously described in the literature as Mercury I or M1) which was characterized by a dark-side periapsis, occurred in March 1973, 146 days after launch. At closest approach, the spacecraft was 700 kilometers above the unilluminated hemisphere. A search for a tenuous neutral atmosphere was conducted during this pass by monitoring the extinction of solar EUV radiation and by observing thermal infra-red emission from a favorable (dark) ground-track. Mariner-10 passed through a region in which the Earth is occulted by Mercury (as viewed from the spacecraft) and this permitted use of a dual-frequency (X- and S-band) radio occultation probe to search for an ionosphere and to measure the radius of the planet. A global magnetic field was unexpectedly discovered in the course of the encounter.

Following a 176 day solar orbit, a second flyby (Mercury II/M2) featured a southern hemisphere passage with a periapsis of ~50,000 kilometers. This trajectory filled a gap in the photographic coverage obtained inbound and outbound during the first encounter. In Section 2.5 is a discussion of the overall coverage achieved and the resolution of the photographs obtained.

During the third, and closest, flyby (Mercury III/M3), the spacecraft flew to within 330 kilometers of the surface, with the primary objective of defining the source of the magnetic field discovered during the first encounter. For this reason M3 like M1 was a dark-side flyby. Because of its closeness to the planet and the absence of an Earth occultation, this pass yielded the most accurate celestial mechanics data obtained during the mission. Partial-frame pictures at the highest resolution (up to 90 m), were acquired near the terminator in areas previously photographed at relatively low resolution during M1.

Data taking continued until March 24, 1975, when, with the supply of attitude-control gas exhausted, the 506 day mission was terminated. The spacecraft was, thereafter, transferred into a retrograde orbit around the Sun, which it still orbits. The total research, development, launch, and support costs for the Mariner series of spacecraft (Mariners 1 through 10) was approximately $554 million and, thus, averaged only $55 million per mission.

2.2   THE MARINER 10 SPACECRAFT

The Mariner 10 bus structure **(Figure 2-2)** was eight-sided and measured approximately 1.4 meters across and 0.5 meters in depth. The weight of the spacecraft was 504 kg, including 80 kg of scientific instrumentation (see Table 2.1) and 20 kg of hydrazine. With its two 2.7 meter by 1 meter solar panels deployed, the span of the spacecraft was 8.0 m. Each panel supported an area of 2.5 m^2 of solar cells attached to the top of the octagonal bus.

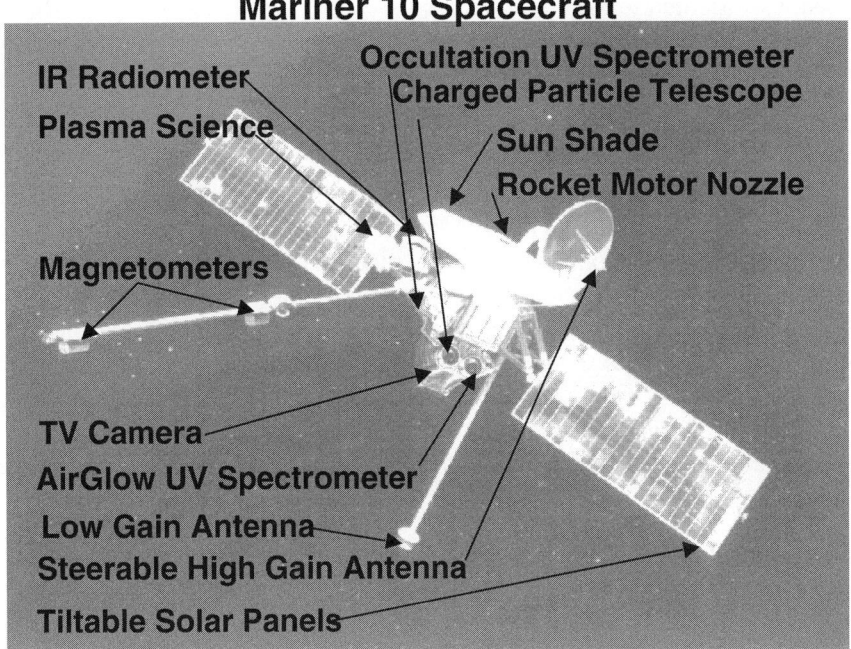

Figure 2-2. Mariner 10 spacecraft, illustrating the spacecraft design described in the text: The instrument package, including cameras boom mounted instruments, including the magnetometer, can be seen, along with the solar panels, later used in the first demonstration of 'solar sailing'.

The spacecraft measured 3.7 m from the top of its low-gain antenna to the bottom of the thrust vector control assembly of its propulsion subsystem. In addition, the high-gain antenna, magnetometer boom, and a boom for the plasma science experiment were attached to the bus. The two degrees-of-freedom scan platform supported two television cameras and the ultraviolet air-glow experiment. A two-channel radiometer was also onboard.

The rocket engine was liquid-fueled and two sets of reaction jets were used to provide 3-axis stabilization. Mariner 10 carried a low-gain omni-directional antenna composed of a 1.4 m wide, honeycomb-disk, parabolic reflector. The antenna was attached to a deployable support boom and driven by two degrees-of-freedom actuators to provide optimum pointing toward the Earth. The spacecraft could transmit at S and X-band frequencies. A Canopus star tracker was located on the upper ring structure of the octagonal satellite and acquisition sun sensors were mounted on the tips of the solar panels.

Simple thermal protection strategies involved: insulating the interior of the spacecraft, top and bottom, using multi-layer thermal blankets and deploying a sunshade after launch to protect the spacecraft on that side which was oriented to the sun.

2.3   THE MARINER 10 SCIENTIFIC PAYLOAD

Table 2-2 lists the instruments and instrument providers for the scientific payload of Mariner 10. The television science and infrared radiometry experiments provided planetary surface data. The plasma science, charged particles, and magnetic field experiments supplied measurements of the interplanetary medium and of the environment close to the planet. The dual-frequency radio science and ultraviolet spectroscopy experiments were designed to detect and measure Mercury's neutral atmosphere and ionosphere. The celestial mechanics experiment provided measurements of the mass characteristics of the planet as well as tests of the theory of General Relativity.

2.4   OVERVIEW OF MARINER 10 OBSERVATIONS

The onboard cameras were equipped with 1500-mm focal length lenses to enable high-resolution pictures to be taken during both the approach and post encounter phases. During the first flyby **(Figure 2-3)**, the closest approach of Mariner 10 to Mercury occurred when the cameras could not photograph its sunlit surface. The imaging sequence was initiated 7 days before the encounter with Mercury when about half of the illuminated disk was visible and the resolution was better than that achievable with Earth-based telescopes. Photography of the planet continued until some 30 min before closest approach, thereby providing a smoothly varying sequence of pictures of increasing resolution. Pictures with resolutions on the order of 2 to 4 km were obtained for both quadratures during M1. Resolution varied greatly, ranging from several hundred kilometers to approximately 100 m. Large-scale features observed at high resolution were used to extrapolate coverage over broad areas photographed at lower resolution. The highest resolution photographs were obtained approximately 30 min prior to and following the darkside periapsis during the first and third encounters. Pictures were taken in a number of spectral bands enabling the determination of regional color differences.

The second (bright side) Mercury encounter provided a more favorable viewing geometry than the first. In order to permit a third encounter it was necessary to target M2 along a south polar trajectory. This allowed unforeshortened views of the south polar region, an area which had not

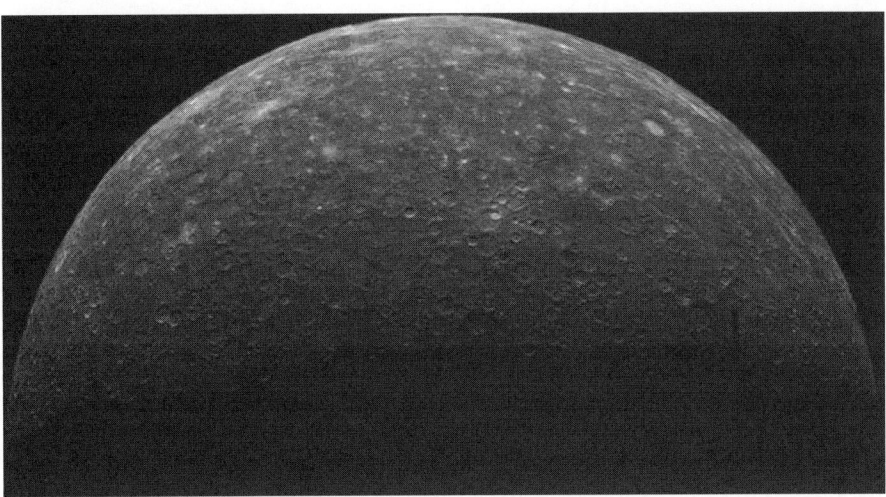

Figure 2-3. Mariner 10 incoming view during the first encounter. (NASA Atlas of Mercury SP432.)

Figure 2-4. Mariner 10 departing view during the third and final encounter. (NASA Atlas of Mercury SP432.)

previously been accessible for study. Images from this region provide a geological and cartographic link between the two sides of Mercury photographed during M1. Stereoscopic coverage of the southern hemisphere was also achieved. Because of the small field of view resulting from the long focal length optics employed, it was necessary to increase the periapsis altitude to about 48,000 km to ensure sufficient overlapping coverage between consecutive images. The resolution of the photographs taken during closest approach ranged from 1 to 3 km.

The third Mercury encounter **(Figure 2-4)** was targeted to optimize the acquisition of magnetic and solar wind data, so that the viewing geometry and hemispheric coverage employed were very similar to those utilized during the first encounter. However, M3 presented an opportunity to provide high-resolution coverage of areas of interest that were previously seen only at relatively low resolution. Because of ground communication problems, the latter pictures were acquired as quarter frames.

Overall, Mariner 10 photographed about 45% of Mercury's surface with a resolution that varied from about 2 km to 100 m (the latter in extremely limited areas).

2.5   MARINER 10 MISSION OBJECTIVES

What was actually accomplished by Mariner 10? The stated objectives of the mission were: (1) primarily, to measure the surface, atmospheric and physical characteristics of Mercury and (2) to measure the atmospheric, surface and physical characteristics of Venus, thereby (3) to complete the survey of the inner planets, as well as (4) to validate the gravity assist trajectory technique, (5) to test the experimental X-band transmitter, and (6) to perform tests of General Relativity theory. We'll describe how well Mariner 10 realized those objectives pertaining to Mercury and advanced the study of that planet in the next four chapters.

2.6   NASA'S ONGOING MESSENGER MISSION

MESSENGER, the MErcury Surface, Space ENvironment, GEochemistry, and Ranging Mission, is a NASA Discovery Mission developed by the Applied Physics Laboratory of the Johns Hopkins University (Gold et al, 2001; Solomon et al, 2001). It was actually launched in August 2004 with modifications to the original scenario **(Figure 2-5)**. After a long seven year cruise, with six challenging gravitational assists (a technique pioneered on the Mariner 10 mission!), including one at the Earth,

Earth Orbit Plane View of MESSENGER's Trajectory to Mercury

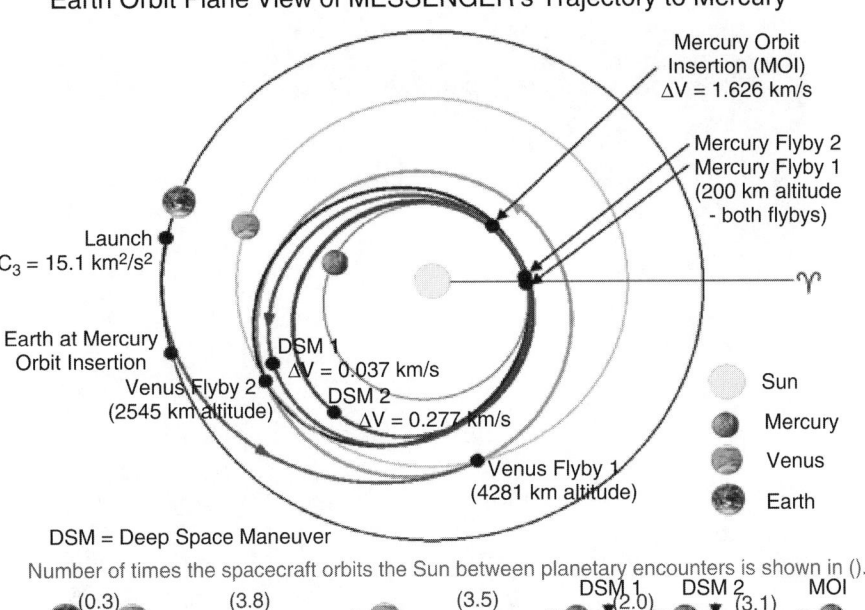

Figure 2-5. Messenger Mission Scenario for original March 2004 launch showing the mission timeline, with extensive use of the gravitational assist technique developed for Mariner 10 during the five year cruise, and the first orbiting of Mercury during the nominal one year orbital mission. (Found at MESSENGER website courtesy of APL.)

two at Venus, and three at Mercury, the spacecraft will undergo orbital insertion, during its fourth encounter with Mercury, in 2011.

The, nearly polar, twelve-hour orbit planned has a high northern latitude periapsis near the terminator. It is highly elliptical, with an altitude that ranges from 200 to 400 km at periapsis to 11,000 km at apoapsis (**Figure 2-5**). Although this configuration will allow 360 degree coverage in the northern hemisphere over the course of the mission, Messenger's orbit, with its high ellipticity and poor illumination at periapsis, is not ideal for spectrometers, which require solar illumination, and will thus provide only low resolution coverage of the southern hemisphere. However, this is the compromise required to enable this state-of-the-art orbital mission to survive in the severe radiation environment of Mercury. The total duration of the mission, four Mercury years, should allow ample opportunity to measure dynamic figure parameters (such as the amplitude of libration) essential in ascertaining the structure of the planet. It is anticipated that 15 Gb of data will be collected during the course of the mission.

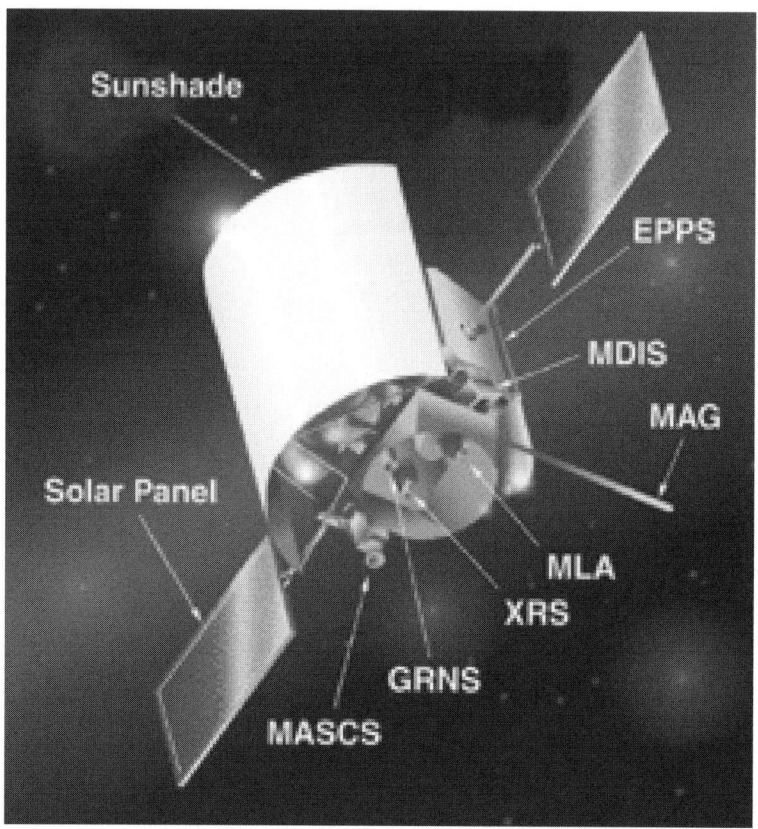

Figure 2-6. MESSENGER spacecraft, illustrating the spacecraft design with the solar panels and sunshade described in the text: The instrument package, including the boom mounted magnetometer, are labeled. The spacecraft is based on a modified NEAR spacecraft design and will use a similar chemical propulsion system. (Found at NSSDC Web site courtesy of NSSDC.)

2.7   THE MESSENGER SPACECRAFT AND PAYLOAD

Messenger (Gold *et al*, 2001) is a fixed body, 3-axis, momentum-controlled spacecraft with chemical propulsion provided by aerojets **(Figure 2-6)**. This design minimizes risk both by eliminating moving parts, thus obviating the chance of mechanical failure, and by exploiting heritage from the NEAR mission. The basic design of NEAR is, however, modified to suit the severe thermal environment at Mercury. The modifications include a

Table 2.3. Payload Flown on Messenger Mission (Gold *et al*, 2001)

Instrument	Range Spec Res	Mss kg	Pwr W	Measurement Spatial Res	BW kbs
MESSENGER Mission, NASA Discovery Mission, Launched August 2004					
Fixed body S/C, 200-440x12000 km, periapsis 60-70N, near terminator, 15 Gb data					
NarrowAngleCamera WideAngle Camera	450-1050 nm 8 color filters	5.5	10	125-250m/pixel B&W >250m BW,>1km color	.4
1 Gamma-ray and 2 Neutron Spctrmtrs	0.1-10 MeV .14 @.6MeV	9	4.5	K,Th,U,Fe,Ti,O ,volatiles Conc to 1m depth, 100's-1000's km/pixel	.1
X-ray Spectrometer	0.7-10 keV 350 ev	4	8	Mg,Al,Si,S,Ca,Fe surface conc, 40-1000's km/pixel	.05
Magnetometer	1024 nT 40 Hz <1-300 sec	3.5	2	Magnetic field, anomalies	.014
Laser Altimeter		5	20	Topo map, 10-50 m spot 100-300 m spacing	.05
UV/Vis Spctrmtr	115-600 nm @<1nm	1.5	1.5	Atmosphere comp,25 km	.9
Vis/IR Spectrograph	300-1450 nm @4nm	1	1.5	Mineralogy maps, 5 km	.9
Energetic Particle and Plasma Spectrometer 1) FIPS and 2) EPS	1)0-10 keV/q 2).01-5MeV/n	2.25	2	High E particles, plasma distribution 1) 360 x 70 2) 160 x 12	.1
Transponder	X-band	5	18	Gravity, interior structure 100's km	.01

lightweight sunshade deployed on the sun-facing side at periapsis, a solar array with optical reflectors, and lighter weight materials.

The Messenger Payload (**Table 2-3**) includes all the instruments that would be expected on a planetary mapping mission, as well as a couple of additional instruments to provide some environmental context. The wide angle camera provides black and white images of the surface with higher average resolutions than Mariner 10 images, as well as far more color information at comparable resolution. The narrow angle camera provides far higher resolution for selected features. The infrared spectrometer provides the first detailed measurements of surface mineral abundances. X-ray, Gamma-ray, and Neutron spectrometers acquire the very first elemental abundance data, for major elements, radioactive elements, and protons (from which water abundance may be inferred) to varying depths in the regolith. The radio science package and altimeter will allow quantitative characterization of the surface and interior morphology. A magnetometer will allow the first comprehensive study of the magnetic field, presumably confirming the presence of the magnetic dipole. Two additional instruments allow characterization of the external environment, including the UV spectrometer to determine the character of the exosphere, and a high energy particle and plasma detector to provide some information on the charged particle environment.

2.8   THE MESSENGER MISSION OBJECTIVES

The relatively poorly constrained yet often surprising results from Mercury have created major controversies about the processes that formed not only that planet but the early solar system itself. Messenger's objectives involve understanding those processes (Solomon et al, 2001). Surface constituent abundances from near IR, X-ray, and Gamma-ray spectrometers on scales ranging from global (bulk) to regional (geochemical province) and occasionally local (stratigraphic unit) will provide insight on solar system formation and Mercury's origin. The high resolution imaging for selected features, combined with comprehensive coverage of the northern hemisphere at higher spectral and spatial resolution than available previously should provide a far greater understanding of Mercury's geological history, and the nature of impact activity in the early solar system. Magnetometer, energetic plasma and particle detector, radio science, and ranging observations should provide insight on the formation and state of Mercury's magnetic core and internal structure. The UV spectrometer and neutron spectrometers should provide measurements which can be used to assess the processes by which Mercury acquired volatiles and an exosphere.

2.9   THE ESA/ISAS PLANNED BEPI COLOMBO MISSION

Bepi Colombo **(Figure 2.7)** is a Cornerstone Mission Concept of the European and Japanese (ESA/ISAS) Space Agencies (Grard et al, 2000; Anselm and Scoon, 2001). At the time of writing, launch is planned for 2014.

Two spacecraft, namely the **Mercury Planetary Orbiter** (MPO) and the **Mercury Magnetospheric Orbiter** (MMO), will be launched, in either split, or single, launch mode. Various options for propulsion are still under consideration. A Solar Electric Propulsion System (SEP) is the likely choice in light of the validation of this technology during the SMART 1 mission (to the Moon). With SEP, both spacecraft will arrive at approximately the same time for capture by Mercury, that is in either less than 2.5 years if they both

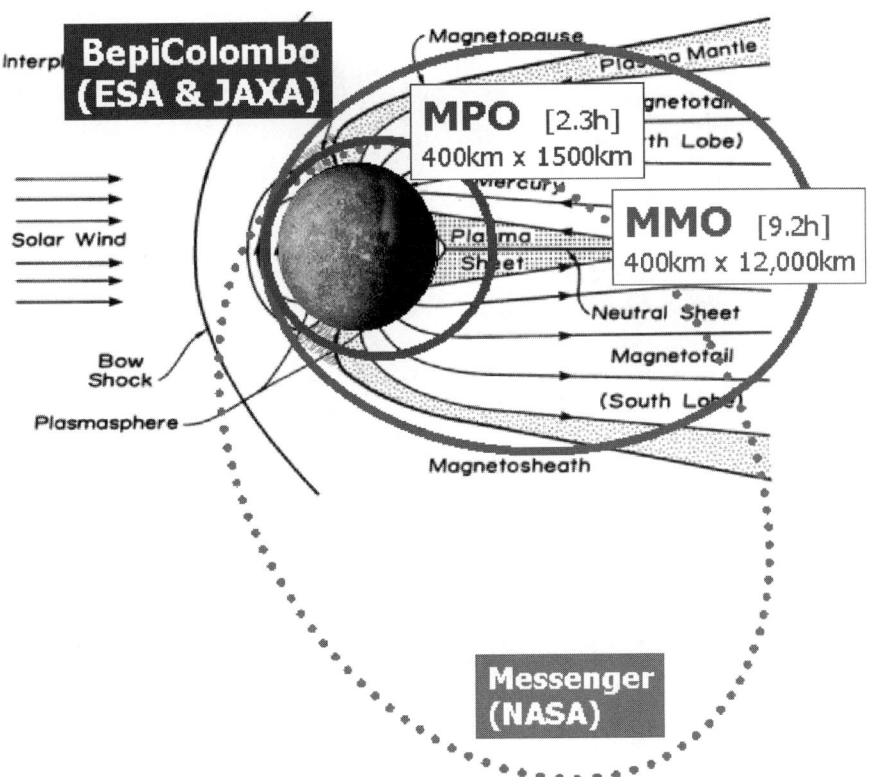

Figure 2-7. Bepi Colombo Mission Scenario for both spacecraft (Grard et al, 2000, ESA Bulletin). This low thurst propulsion system trajectory will require a 2.5 to 3.5 year cruise. Mercury will then be studied from orbit over a one year period. (Courtesy of ESA)

go directly to the planet or in 3.5 years if a gravitational assist strategy is used. The plan is to insert both spacecraft, probably using chemical propulsion systems, into nearly polar orbits and have equatorial periapses, thereby allowing 360^0 coverage of the entire planet to be achieved during the lifetime of the mission, which is nominally one year, as in the case of Messenger. Initially, both MPO and MMO will be inserted into an anti-solar, equatorial periapsis, and an elliptical 400 x 1,500 km orbit. The selection of periapsis in the anti-solar hemisphere is a strategy adopted to deal with the extreme radiation environment, but the trade off will be lowering the best available spatial resolution. Gradually, the MMO will be inserted into a resonant orbit, while maintaining an anti-solar, equatorial periapsis, to attain an ellipticity of 400 x 12,000 km (which is desirable for a magnetospheric mission). The MPO is expected to collect over 1500 Gb of data and the MMO 1.5 Gb of data during the nominal mission.

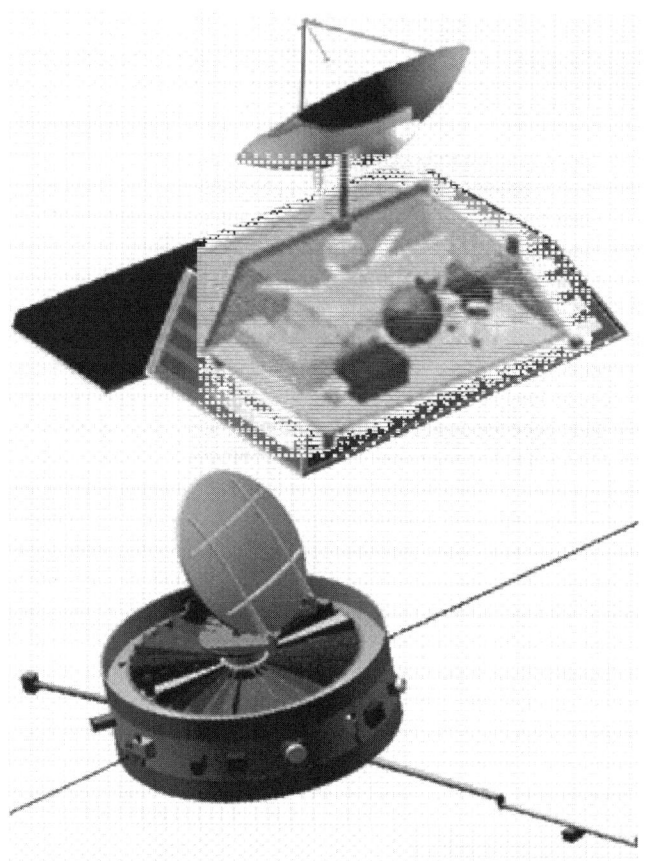

Figure 2-8. Bepi Colombo ESA MPO and ISAS MMO (Grard et al, 2000, ESA Bulletin). These views illustrate the spacecraft design concept, with solar panels and sunshade incorporated into the design. The MPO (Mercury Planetary Orbiter) (above with cutaway insert) is a 3-axis stabilized, nadir pointing spacecraft. The MMO (Mercury Magnetospheric Orbiter) (below) is a spin stabilized spacecraft with spin axis perpendicular to the ecliptic plane. Both spacecraft use solar electric propulsion to get to Mercury, and a chemical propulsion system for orbital insertion and all activities which follow. (Courtesy of ESA)

2.10   THE BEPI COLOMBO SPACECRAFT AND PAYLOAD

Bepi Colombo is a dual platform mission **(Figure 2.8)**. The MPO, to be provided by ESA, features a nadir-pointing, 3-axis stabilized design, which is optimal for a mapping mission. It is designed to provide close range studies of the surface and, from its measurements, the internal state and dynamic figure properties of Mercury can be derived. The MMO, to be provided by ISAS, features the spin stabilization and 15 rpm spin rate optimal for a magnetospheric mission. It is designed to provide information on the wave and particle environment surrounding Mercury. The two payloads **(Table 2.4)** are configured to provide complementary measurements of the planet in the context of its external environment.

The MPO will include a traditional planetology instrument suite. Wide and narrow angle cameras provide average 200 m/pixel resolution for the entire surface and 20 m/pixel resolution for selected features, This is comparable to what is provided by Messenger but, in the case of Bepi Colombo, coverage of the southern hemisphere is additionally available. The infrared spectrometer provides significant improvements with respect to spectral coverage and spatial resolution relative to MM, thereby allowing a more detailed study to be made of local variations, particularly in the southern hemisphere. The Gamma-ray and Neutron spectrometers have comparable performances to MM. The CIXS X-ray spectrometer, however, will be capable of higher sensitivity, as well as of higher spectral and spatial resolution than is provided by the proportional counters used on Messenger. This performance will greatly facilitate the direct comparison of mineralogical and major elemental abundances of elements including iron, for features on the scale of 1 to 2 km in size. The Ultraviolet spectrometer is comparable to the one flown on Messenger and should allow a study to be made of temporal variations in atmospheric constituents, and atmospheric dynamics in the southern hemisphere. Also, the radio science package and laser altimeter are comparable to the ones flown on Messenger, but, again, coverage of the southern hemisphere should result in better global modeling of Mercury's interior structure.

The MMO will include a traditional magnetospheric instrument suite. The magnetometer is comparable to the Messenger instrument. The combined charged particle detectors will provide a more comprehensive survey than is available from Messenger of the nature of charged particle behavior in the magnetosphere. The search coil and electric antenna will allow detection of local emission sources. The camera will provide both additional and complementary imaging information.

2.11   THE BEPI COLOMBO MISSION OBJECTIVES

The objectives of Bepi Colombo focus on obtaining the kind of coverage necessary to fill in particular gaps in our knowledge of planet Mercury (Grard *et al*, 2000). Determining the nature of the unimaged hemisphere and polar deposits as well as composition of the entire surface

Table 2-4. **Proposed Payload for Bepi Colombo**

ESA Mercury Planetary Orbiter (MPO)

http://sci.esa.int/science-e/www/object/index.cfm?fobjectid=36098 as of 1/1/2005

3axis, nadir pointing S/C, orbit 400x1500 km, periapsis anti-solar point, 1500 Gb data

Instrument	Range	Mass kg	Pwr W	Measurement	Bw kbs
HiRes Stereo Cameras Vis/IR Spectrometers (SIMBIO-SYS)	350-1050 nm	12	16	20m/pixel narrow 200m/pixel wide	40
IR Spectrometer (MERTIS-TIS)	.8-2.8 u 128 channels	6	10	Mineralogy maps .15-1.5 km resolution	1.5
UV Spectrometer (PHEBUS)	70-350 nm	3.5	3	Atmosphere comp	2
X-ray Spectrometer (MIXS) and Solar Monitor (SIXS)	0.5-10 keV	4.5	8	Na,Mg,Al,Si,S,Ca,Fe surface conc at 10's km monitor Xray source(Sun)	0.1
Combined Gamma-ray Spectrometer and Neutron Spectrometer (MGNS or MANGA)	0.1-8 MeV 0.01-5 MeV	7.5 5	5 3	K,Th,U,O,Fe conc to 10s cm depth at 100's km Volatiles to 1 meter at 100's km	0.1 0.05
Laser Altimeter (BELA)				Topography map	
Radio Science Package (MORE)	32-34 GHz 10^{-4}-10^{-1} Hz	11.5	15.3	Gravity, interior structure	0.1
Magnetometer (MERMAG)	4096 nT	0.88	0.35	Mag Field, anomalies	0.8
Neutral/Ionized Particle Analyzer (SERENA)				space (Particle-induced) weathering processes	

ISAS Mercury Magnetic Orbiter (MMO)

http://www.stp.isas.jaxa.jp/mercury/pro-mmo.html#AO as of 1/1/2005

15 rpm spinstabilized S/C, orbit 400x12000 km, periapsis anti-solar point, 1.5Gb data

Instrument	Range	Mass kg	Pwr W	Measurement	Bw kbs
Magnetometer (MGF)	4096 nT	0.88	0.35	Mag field, anomalies	0.8
Low/Hi Energy Ions, Electrons, Hi Energy Energetic Neutrals (MPPE)	.0-300 keV	7.5	10	electron, Ion, neutrals mass, charge, energy distributions	1.0
Electric Field, Radio Wave, Plasma Wave Detectors (PWI)	0.1-1 MHz	1.0	5	electric field, radio wave, and plasma wave sources	0.5
Na Imaging (MSASI)				Na spatial and temporal distribution	
Dust Detector (MDM)				Dust count and moment	

are of high priority. The onboard spectrometers with broader and more sensitive coverage than previously available should provide this information. Combining the more sensitive spectrometer results with the magnetometer data should lead to a more comprehensive understanding of how iron, the cause of Mercury's high density, is distributed, and thereby increase our understanding of core formation and the evolution of the magnetic field. Combining the more sensitive spectrometer results with more extensive higher resolution images should lead to new insight into the nature of geological evolution. Combining the more sensitive spectrometer results with altimeter and radio science package data should allow the interior structure of the planet to be better understood. The ultraviolet spectrometer should allow further characterization of the exosphere, and the entire MMO package will permit a comprehensive study to be made of local interactions between particles, waves, plasmas, and the solar wind in the magnetosphere in the absence of an ionosphere.

2.12   SUMMARY

The Mariner 10 mission provided the basis for the basis for our current understanding of Mercury but provided some startling revelations and many unanswered questions. The MESSENGER mission now enroute to Mercury is a NASA Discovery Class planetary orbiter which should provide the basis for understanding Mercury's magnetic field, the first direct compositional data for Mercury as well as in situ observations which will provide intriguing snapshots of the environment around Mercury. Bepi Colombo, an ESA multi-platform mission that could potentially provide more information on the dynamic environment is currently being planned.

2.13   REFERENCES

Anselm, A. and G. Scoon, Bepi-Colombo, ESA Mercury/Cornerstone Mission, *Planet. Space Sci.* **49**, 409-420, 2001.

Gold, R.E., S.C. Solomon, R.L. McNutt, A.G. Santo, J.B. Abshire, M.H. Acuna, R.S. Afzal, B.J. Anderson, G.B. Andrews, P.D. Bedin, J. Cain, A.F. Cheng, L.G. Evans, W.C. Feldman, R.B. Follas, G. Gloeckler, J.O. Goldstein, S.E. Hawkins, N.R. Izenberg, S.E. Jaskulek, E.A. Ketchum, M.R. Lankton, D.A. Lohr, B.H. Mauk, W.E. McClintock, S.L. Murchie, C.E. Schlemn III. D.E. Smith, R.D. Starr and T.H. Zurbuchen, The MESSENGER mission to Mercury: scientific payload, *Planet. Space Sci.* **49**, 1467-1479, 2001.

Grard, R., M. Novara. And G. Scoon, BepiColombo: an interdisciplinary mission to a hot planet *ESA Bul.* No. **103**, 10–19, 2000.

Solomon, S.C., R.L. McNutt, R.E. Gold, M.H. Acuna, D.N. Baker, W.N. Boynton, C.R. Chapman, A.H. Cheng, G. Gloeckler, J.W. Head, S.M. Krimigis, W.E. McClintock, S.L. Murchie, S.J. Peale, R.J. Phillips, M.S.

Robinson, J.A. Slavin, D.E. Smith, R.C.G. Strom, J.I. Trombka and M.T. Zuber, The MESSENGER mission to Mercury: scientific objectives and implementation, *Planet. Space Sci.* **49**, 1445-1465, 2001.

2.14 SOME QUESTION FOR DISCUSSION

1. What surprises and unresolved issues did the Mariner 10 mission leave in its wake?

2. Compare the MESSENGER and Bepi Colombo missions. How would you improve either mission and what would be the impact on spacecraft resources (mass, power, bandwidth) and cost.

3. Discuss what spatial and temporal resolutions and coverage will be provided by instruments on both missions, and the adequacy and limitations of this instrumentation in revealing the dynamic character of Mercury's interaction with its environment.

Chapter 3

MERCURY'S INTERIOR

3.1   PRESENT UNDERSTANDING OF MERCURY'S INTERIOR

Because Mercury has a striking cluster of unique characteristics (Chapter 1), the planet is in a position to contribute important inputs for models of Solar System formation, evolution, and structure. Two of the most fundamental problems in planetary science, the formation of the solar system and the dynamo generation of magnetic fields in planetary interiors, can be addressed by studying Mercury's deep interior. Yet, Mercury is the terrestrial planet for which we have the most limited knowledge of geology, geophysics, and geochemistry, and thus the least direct knowledge of the interior. We do know that the planet is the smallest in mass, closest to the sun, and greatest in mean uncompressed density (5.3 g/cm^3 at 10 kbar) among the terrestrial planets. In all these ways, Mercury thus has a unique end-member status, allowing it to provide previously unavailable, yet essential, inputs for understanding the solar system.

3.2   BULK PROPERTIES

The bulk properties of Mercury are summarized in the context of the other terrestrial planets in **Table 3.1.**

Combined ground-based (Lyttleton, 1980, 1981; Branham, 1994) and Mariner 10 measurements (Anderson et al, 1987) have established that, although the mass of Mercury is small (3.3 x 10^{26} g) relative to the masses of the other terrestrial planets, it is quite massive in relation to its size (4878 km). The average uncompressed density of Mercury (5.3 g/cm^3), presently known indirectly through dividing the measured mass by the volume, is much larger than the Earth's uncompressed density (4.1 g/cm$^{3)}$). This implies that Mercury's interior is 60% FeNi alloy by volume (Lewis and Prinn, 1984; Cook, 1982).

Table 3-1. **Extreme Properties of End-Member Mercury**

Planet	Mercury	Earth	Mars
Diameter, km	4880	12756	6592
Mass, g	$3.3*10^{26}$	$6.0*10^{27}$	$6.4*10^{26}$
Density uncompressed g/cm^3	5.4	5.5	3.9
Bulk Composition in terms of iron and magnesium	Fe>>FeO? FeO<MgO? less volatiles? Fe$_{mercury}$<Fe$_{Earth}$?	FeO>MgO crust FeO>>Fe crust/mantle Fe>>FeO core	FeO>>Fe FeO>MgO more volatiles Fe$_{mars}$<Fe$_{Earth}$
Crust Composition	Little Fe silicate	ocean basalt/continent granite	andesite basalt
Core by Volume	42%	16%	9%
MagneticDipoleMoment J/T	4.9×10^{19}	7.9×10^{22}	$<2.1 \times 10^{18}$
Internal Heat Flow (Hauck et al, 2004) (Hagermann, 2005)	Modeled >2000K >15 mW/m^2	5200-6200K 25-150 mW/m^2	Modeled 2970-4170K 32 mW/m^2

What does Mercury's extreme density imply about its interior structure and composition? It implies that the planet is composed, to a large extent, of heavy elements, particularly iron, the most abundant heavy element in the solar system. Although no direct iron abundance measurements have been made, Mercury's high density implies a high bulk abundance of iron and a metal to silicate ratio twice that of any other terrestrial planet or the Eucrite parent body (Cameron et al, 1987). If iron is concentrated in Mercury's core, as Mariner 10 magnetic field measurements apparently indicate, then the planet must have a huge iron-rich core (75% of the planet's diameter) and a relatively thin (600 km) combined crust and mantle, compared to the Earth (2900 km). This implies that, with Mercury's much larger high density core, it must have a lower density mantle than the Earth's, perhaps consisting of lower pressure forms of olivine and pyroxene (Cook, 1982). **Figure 3.1** illustrates the large size of Mercury's core relative to the other terrestrial planets.

3.3 MAGNETIC FIELD AND CORE FORMATION

One of the most important Mariner 10 discoveries was made when the spacecraft passed nearly directly above the rotational north pole of Mercury (at an altitude of 327 km), and measured a magnetic field strength of up to 400 nT (Ness et al, 1974; 1975). The variation and magnitude of the field along the spacecraft trajectory suggested a planetary field of internal origin,

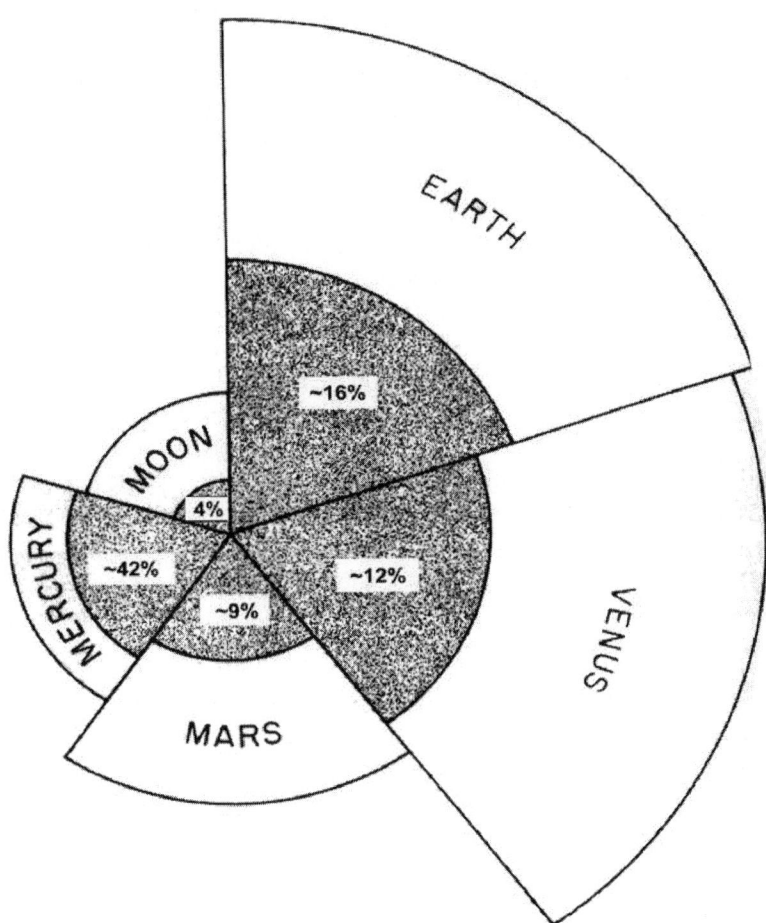

Figure 3-1. Relative sizes of Terrestrial Planets and Their Cores by volume showing Mercury's disproportionately large core (Strom, 1987 (Courtesy of Smithsonian Institution).

closely approximated by a dipole aligned with the rotation axis deep within the planet. (See also Chapter 6, the Magnetosphere Section.) The inferred magnetic dipole moment of up to 6000 nT m^3 (~5x10^{-4} that of the Earth's), endows Mercury with the weakest global intrinsic magnetic field (Christon, 1987; Russell et al, 1988) of any magnetized planet, a field that is 100 times weaker than the Earth's.

If Mercury's field is due to dynamo generation in a fluid core, it is the weakest such dynamo known; thus, not long after its discovery, some workers (Ness et al, 1975; Cassen et al, 1976) proposed that it was a fossil magnetic field from an ancient, inactive dynamo. Others argued that, although weak, the field appears to be too strong to be explained by remnant magnetism (Schubert et al, 1988). Better field characterization is required to test this, and other more exotic, possibilities (e.g. Stevenson's 1987 thermo-electric 'dynamo'). One major difficulty with the existence of a still active dynamo, discussed in detail in the next section, is the requirement for a partially molten core, and the retention of heat in the mantle, implying a longer geological history than demonstrated by other evidence.

Several attempts were made to model the magnetic field configuration using Mariner 10 measurements (e.g. invoking a quadrupole or putative dipole offset). Connerney and Ness (1988) demonstrated that all such attempts lead to non-unique models (a single spacecraft over an essentially non-rotating body results in a poorly constrained inverse problem).

More recently, the magnetometer aboard the Global Surveyor Spacecraft recorded intrinsic, intense magnetization at Mars that was mainly confined to the heavily cratered, ancient, southern highlands (maximum strength 220 nT at a mapping altitude of 370-438 km) (Connerney et al, 2001). This discovery prompted several researchers to return to the Mariner 10 measurements to see if, rather than being attributed to the dipolar configuration discussed above, these observations might be interpreted to indicate the presence of irregularly distributed, strong, remnant fields on Mercury's surface. The Mariner 10 data were found to be inadequate to allow a distinction to be made between these possibilities.

3.4   STRUCTURE OF MERCURY'S CORE

Measurements suggesting that Mercury has an appreciable, intrinsic, magnetic field, have consequences for the possible state of its interior. Evidence for an internal dynamo on Mercury, whether still active or not, is certainly evidence for global differentiation which produced an Earthlike interior structure early in Mercury's history. The observed tectonics (Melosh and McKinnon, 1988) that led to widespread volcanism and scarp formation appear to have resulted from early core formation. The prevailing understanding concerning planetary, dipolar, magnetic fields is that they are generated by electrical currents, induced by dynamo action in a thermally convecting, differentially rotating, out liquid metallic core split from the inner core and rotating fastest nearest the center (Campbell, 1997). A complex field is produced, that can be approximated as a dipole near the surface. If the field is indeed generated by a dynamo, then some part of the iron-rich, electrically-conducting, core must remain fluid today. However, many models of the composition and thermal evolution of Mercury indicate that a differentiated Mercury would have cooled and solidified long since (Sigfried and Solomon, 1974; Cassen et al, 1976; Fricker et al, 1976). The age and global distribution of the compressional scarp system also implies shrinking and core solidification long ago (Cassen et al, 1976).

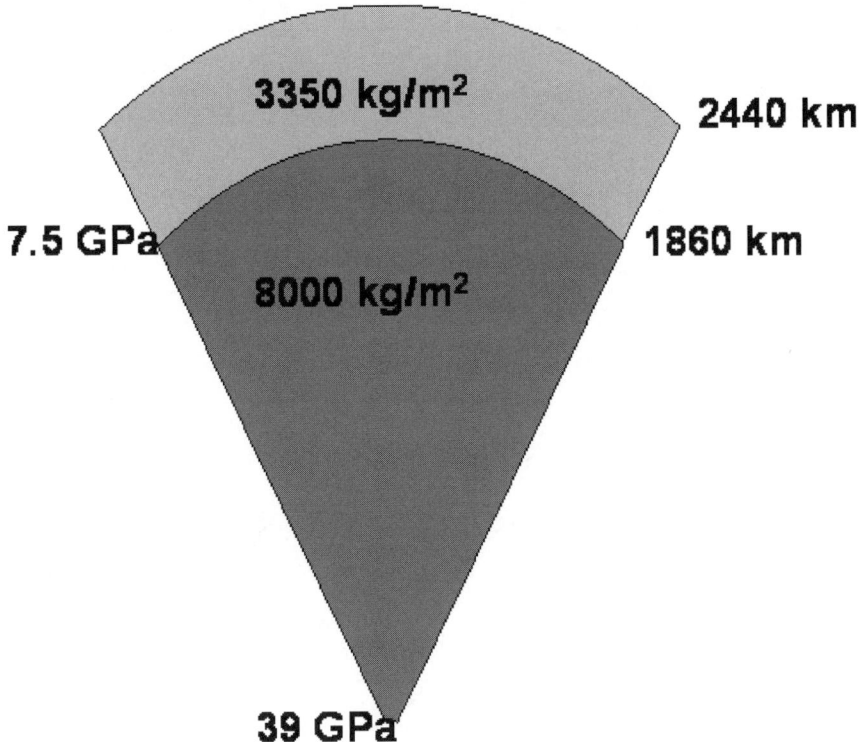

Figure 3-2. Simple model of Mercury's interior structure: Two-layer model of Mercury's interior structure based on observed planetary mass and radius and best inferred core density. (Reprinted from Spohn et al, 2001, with permission from Elsevier.)

A typical simple 'model' of core structure is shown in **Figure 3.2** (Spohn et al, 2001). As a result of a large iron-rich core to account for its observed density, the mantle and crust should be a thin silicate shell, with pressure, density, temperature, and elastic moduli varying little (Spohn et al, 2001). However, evidence that Mercury has an appreciable intrinsic magnetic field has a direct bearing on the current state of its interior and its origin.

If the Hermean magnetic field is really produced by dynamo action, at least part of the iron-rich, electrically-conducting core must have avoided solidification and presently be in a fluid state. It is possible that high internal temperatures could have been maintained by adding material with lower thermal diffusivity to the mantle, by raising the viscosity through the loss of volatiles, or by adding radioactive isotopes or lighter elements, such as silicon or sulfur, to form a lower melting point alloy in the core. There are difficulties with long-lived radionuclides as the heat source unless they are present in densities greater than in the Earth's core (Cassen et al, 1976). Sulfur (Stevenson et al, 1980) or silicon in a highly reduced form, as found in enstatite (Keil, 1968), provide the most reasonable candidates for an

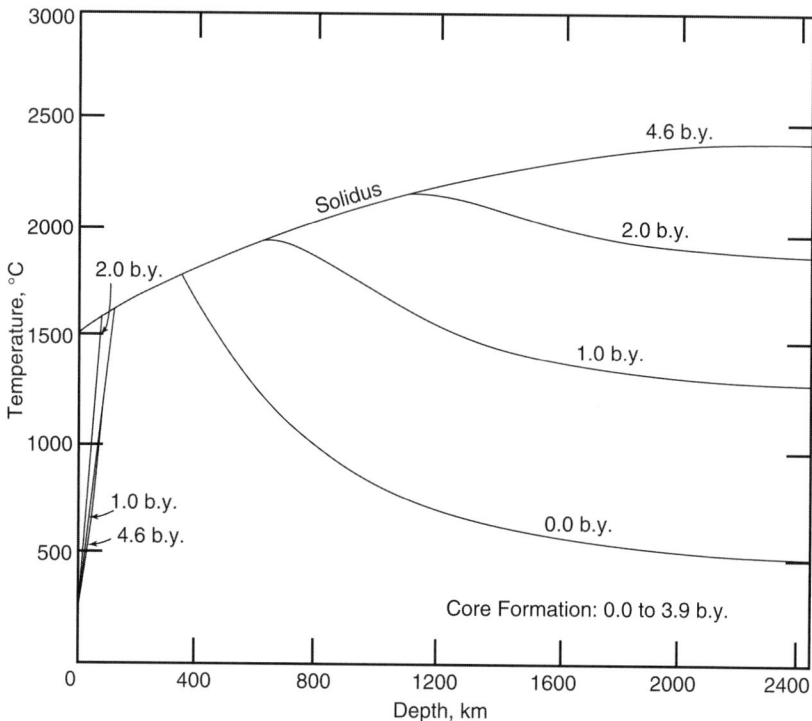

Figure 3-3. Thermal history of Mercury, based on equilibrium condensation origin and initially molten crust so that initial temperatures highest are highest near the surface. (Reprinted from Solomon, 1976, with permission from Elsevier.)

alloying element which would lower the melting temperature. The density, size, and evolution of the core would depend on the nature and abundance of such an element, which could be deduced from a more comprehensive understanding of the planet's tectonic history. The addition of a few percent sulfur would allow a liquid outer core of over 1000 km thick to remain at the present epoch. An entirely liquid core would result with 7% sulfur (Schubert et al, 1988; Okuchi, 1997). Tidal (Cassen et al, 1976; Schubert et al, 1988) or gravitational (Solomon, 1976) heating have also been proposed as mechanisms for keeping an outer core molten. By starting with an initially hot planet with an interior approaching the equilibrium black body temperature at Mercury's present distance from the sun, and allowing metal-silicate differentiation to proceed slowly downward by gravitational infall, the distributed gravitational heating model (**Figure 3.3**) produced 1) outer layers which were rapidly resurfaced as well as 2) a core remaining in a partially molten state after 4.6 billion years. Although Gubbins (1977) has calculated that a dynamo is possible at the low spin rate of Mercury, if Mercury's magnetic field is attributed to dynamo generation in a fluid core,

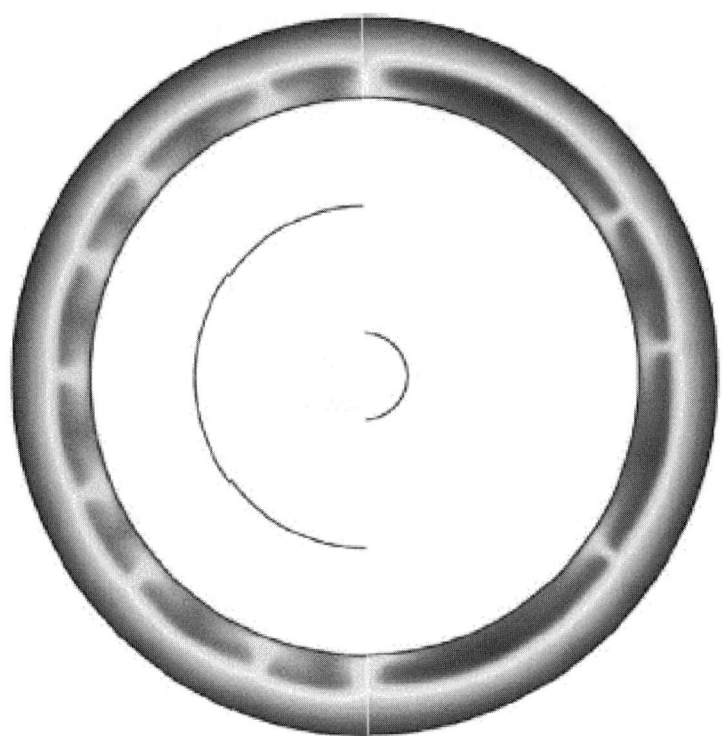

Figure 3-4. Thermoelectric core models: Convection cells, temperature, and inner core sizes derived for two thermal evolution calculations using 2D convection code for thermoelectric core model (Conzelmann, 1999). On the left, sulfur concentration in the core is 2% and the viscosity is simply temperature dependent. On the right, when sulfur concentration decreases to 0.1%, a higher activation energy is required due to increasing pressure. (Reprinted from Spohn et al, 2001, with permission from Elsevier.)

then it is the weakest dynamo known (Christon, 1987; Russell et al, 1988).

Both Stevenson (1975) and Solomon (1976) proposed that Mercury's magnetic field could be the result of remnant magnetism surviving from an ancient dynamo field. Stevenson (1987) related Mercury's magnetic field to a thermo-electric dynamo in which current flow is driven by temperature differences at an irregular core/mantle boundary, thereby generating helical convective motions in the outer layer of the core to produce the modest field measured today **(Figure 3.4)** (Spohn et al, 2001). Lateral pressure gradients applied to the core/mantle boundary due to convection currents, would in this scenario give rise to irregularities comprising undulations of the order of a few kilometers, and these in turn would produce lateral temperature gradients associated with electrical potential differences. The wavelength of a particular undulation is related to the wavelength of the convection pattern. Similar undulations could also occur at the crust/mantle boundary and, in some cases, these would be difficult to distinguish from core/mantle features. Better field characterization is presently required to support the thermo-electric, or alternative dynamo mechanisms.

3.5   SHAPE, GRAVITY FIELD, AND INTERNAL STRUCTURE OF MERCURY

Knowledge of Mercury's internal structure could allow theories of magnetic field generation and of thermal history to be constrained. Such knowledge could also constrain the initial rotation state. How well is Mercury's internal structure known?

Although Mercury's internal structure is not well constrained, Anderson and coworkers (1996) have used radar observations to derive information concerning Mercury's equatorial shape and concluded that the center of mass of the planet is offset in the equatorial plane. This suggests the existence of an asymmetry in the crustal thickness of Mercury comparable to that known to exist on the Moon.

Certainly, gravity measurements cannot be used to determine an internal structure uniquely. However, low degree gravity parameters combined with dynamic figure parameters, such as rotation state, can provide constraints on the interior structure by determining the moments of inertia of the principal axes. If a liquid layer, such as an outer core, decouples the inner core from the crust, then the moment of inertia of the crust may be determined separately. It is possible to determine anomalies, or lateral inhomogeneities in the mantle, on a scale comparable to the altitude of the orbiter. The depth of the mantle and core can be constrained using gravity and topography data (from overlapping images) under the assumption that local variations are caused by variations in crustal thickness.

The gravity field of Mercury was poorly known even after Anderson and coworkers (1987) reanalyzed the Mariner 10 flybys to give improved estimates of second degree coefficients associated with principal axes. Available measurements of the moments of inertia (MoIs) from Mariner 10 (C_{20} with a 30% accuracy and C_{22} with a 50% accuracy) were not sufficiently accurate to distinguish between a differentiated and a homogeneous body, thereby providing little guidance for compositional and thermal models of the interior. (Anderson et al, 1987). These coefficients must be determined with a 10% accuracy to unambiguously model the state of the core (Peale, 1988) and the internal magnetic field structure. The existence of the outer liquid core must be verified and its properties constrained by deriving the amplitude variations of Mercury's forced physical libration. Such variations could in principle be derived by determining moments of inertia which appear in expressions for second-degree coefficients of the planetary gravity field.

3.6   SEARCH FOR A LIQUID CORE/SHELL

Observations that can determine the existence of a fully liquid core, or the extent of a liquid shell about the core, in Mercury's interior were described by Peale (1976, 1988) who demonstrated that measurement of the obliquity of the lowest degree gravitational harmonic coefficients and of the physical librations about the commensurate spin rate allow the extent of any liquid component to be identified. For the experiment to succeed, it is necessary that the core does not follow the 88 day physical libration cycle of the mantle but rather follows the mantle on the time scale of the 250×10^3 year precession of the spin.

Mantle libration about the mean resonant angular velocity arises from the periodically reversing torque on the planet as Mercury rotates relative to the Sun. The amplitude of this libration (ϕ_o) is approximately equal to $(B-A)/C_m$ (where A and B are the two equatorial principal moments of inertia of the planet and C_m is the moment of the solid outer parts of the planet about the rotation axis).

Dissipative processes act to bring Mercury to rotational Cassini state 1, with obliquity (θ) close to $0°$, thereby yielding a relationship between θ, the moments of inertia, and other orbital parameters (Peale, 1988). The differences in the moments of inertia also appear in expressions for the second degree coefficients of the planetary gravity field, C_{20} and C_{22}. In Cassini State 1, the spin vector remains coplanar with the orbit normal and the normal to the plane of the Solar System about which the orbit precesses.

This relationship offers a means to determine the possible presence of a fluid outer core and its associated radius through measuring the second-degree gravity field, the obliquity (θ) and the physical libration amplitude (ϕ_o). The ratio of these moments of inertia, the solid outer parts of the planet to that of the planet as a whole (C_m/C), can be used to determine whether the planet is solid or partially liquid, as expressed in ***Equation 3.1***.

$$C_m/C = (C_m/(B-A)) (B-A/MR^2) (MR^2/C) \leq 1$$

The first quantity in brackets is equivalent to ϕ_o. The second is equal to C_{22} and the third is obtained from the relationship between θ and the second degree gravity field coefficients. Thus, C/MR^2 can be derived to an accuracy that is limited by the uncertainty in θ, while C_m/C is determined to an accuracy that is principally limited by the uncertainties in ϕ_o and θ.

If the ratio C_m/C is equal to 1, then the core of Mercury is solid. If the ratio is between 1.0 and 0.5, the core is partially to fully molten. For values of C_m/C less than 1, the radius, Rc, of the fluid outer core is determined. If C_m/C is equal to 0.5 for Rc/R equal to 0.75, then, from C/MR^2, the radius of any solid inner core may be estimated or bounded.

Recently, a series of high precision rotation measurements made by using Earth-based radar observations have revealed that libration amplitudes are three times larger than they would be for a solid core coupled to the mantle (Margot et al, 2004; Peale, 2004). These data provide direct evidence that Mercury's core is at least partially liquid and decoupled from the outer solid mantle and crust, as well as further observational support for a dynamo generated magnetic field. Now, the interior structure of Mercury is no longer merely implied on the basis of the mass and size, and thus the density, of the planet, but constrained to be at least partially molten, with profound implications for Mercury's geological history, as described in detail in Chapter 4, the Surface section.

3.7   SOLAR SYSTEM FORMATION

Explanations for Mercury's origin and means of acquiring so much iron remain highly speculative. The models, described in detail in Chapter 1, generally begin with the following scenario for solar system formation. The Sun formed at the hot center of a large cloud of gas and dust (the Solar Nebula) which became unstable and collapsed under the force of gravity. The composition of solid particles (planetesimals) from which planets eventually formed would have exhibited a gradient based on temperature between the inner and outer portions of the solar nebula.

3.8   EQUILIBRIUM CONDENSATION MODEL

One of the oldest and simplest models for the origin of solar system bodies relies exclusively on the mechanism of chemical **equilibrium condensation** and predicts composition based solely on a body's heliocentric distance, or provenance, and thus its formation temperature and pressure in the Solar Nebula (Lewis, 1972, 1988). This model assumes that a planet's original and present day provenances are one and the same, and thus it is composed of planetesimals of that provenance. The terrestrial Planets would have formed relatively close to the Sun and be characterized by high abundances of non volatile elements (such as silicon, aluminum, magnesium and iron), while the Outer Planets would show large abundances of volatile elements (such as hydrogen and helium). Here, the high density of Mercury was attributed to the slightly higher condensation temperature of iron as compared with the condensation temperatures of the magnesium silicates present in the cooling Solar Nebula. In particular, at Mercury's distance from the proto-sun, the ratio of solid metal to silicate was derived to be much higher than in the formation zones of the other Terrestrial Planets

(Lewis, 1972). While the Lewis model (1972) predicts enrichment of metallic iron, it still predicts far too low an uncompressed density of ~4.5 g/cm³ for Mercury. It predicted severe depletion of sulfur, which has been considered a prime candidate for maintaining a partially molten core, and of volatiles, which apparently have been observed at Mercury's poles.

Modified planet formation models (Lewis, 1988) where 10-40% of the planet accreted from planetesimals perturbed to cross Mercury's orbit, predict a nominal amount of FeS which supplies the necessary sulfur, but fall far short of providing enough iron to account for the observed core density. Several different hypotheses have been proposed to account for the large discrepancies between prediction (relative to the equilibrium condensation model) (Lewis, 1988) and observation regarding Mercury's formation. The validity of the models could be assessed by measuring the bulk abundance and variations in the major/minor elements Na, Mg, Al, Si, and Fe along with the refractory (ref), alkali (alk), residual (rsd), and Fe-bearing mineralogical components (**Table 3.2, Figure 3.5**) (Lewis, 1988). Although a correlation between final heliocentric distance and planetesimal composition remains, the importance of primordial chemical fractionation processes is reduced in these models.

Table 3-2. **Formation Model Implications for Elemental Abundances**

Model	Elements					
	Mg	Fe	Si	Na (alk)	Al (ref)	Th (rsd)
Equilibrium Condensation	H	LL	M	L	L	H
Selective Accretion	ML	ML	L	H	MH	H
T Tauri Vaporization	M	ML	L	L	HH	HH
Giant Impact	ML	MH	L	M	L	L
Reduced/Oxidized	L	H	M	L	L	L
Enstatite Chondrite	H	H	H	H	L	L
H = High; M = Moderate; L = Low						

A range of processes, including dynamic or physical sorting (Weidenschilling, 1978; Cameron et al, 1987), repeated impacts (Wetherill, 1988, Benz et al, 1988), or different initial starting compositions (Wasson, 1988) have been invoked to deal with these discrepancies and to bring Mercury's theoretical density closer to observed values. Weidenschilling (1978) suggested aerodynamic sorting through gas drag on metals and silicates in the solar nebular during accretion enhanced the Fe/Si ratio relative to the strict equilibrium condensation model. Consideration of these models does not rule out the possibility of a hybrid explanation, where the initial composition was based on one mechanism, such as equilibrium condensation controlled by temperature and thus heliocentric distance, and further fractionation occurred through another.

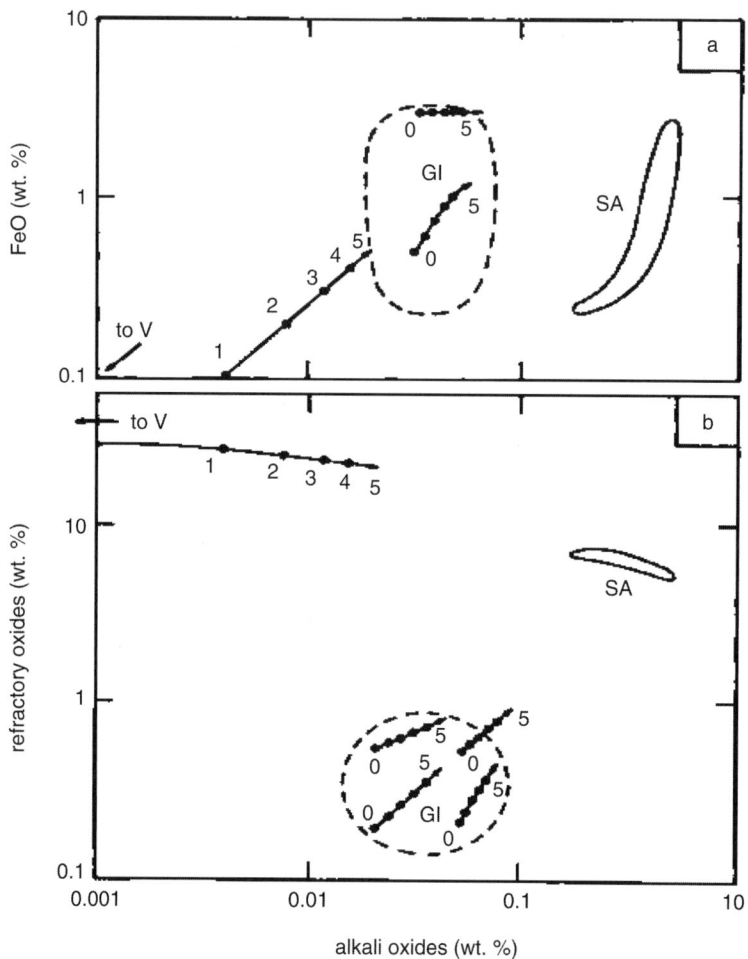

Figure 3-5. Comparison of Predicted Bulk Compositions of Mercury, based on FeO, alkali and refractory oxide content, predicted by three different models, Selective Accretion (SA) from Weidenschilling, Giant Impact (GI), and residual mantle from volatilization (V).Arrows indicate effect of additional average chondritic material from late infall (From Lewis, 1988, Origin and Composition of Mercury, in Mercury, Copyright 1988, The Arizona Board of Regents. Reprinted by permission of the University of Arizona Press.)

It should be mentioned that, based on trends in anomalies in Mercury's orbital elements, Van Flandern and Harrington (1976) have joined others in proposing that Mercury originated as an escaped satellite of Venus. However interesting this model may be on the basis of dynamic considerations, it in no way solves the problem of Mercury's iron abundance. In fact, Ward and coworkers (1976) were able to account for many of these anomalies on the basis of solar tidal spindown. A number of workers (Burns, 1976; Peale, 1976) have predicted that this spindown would create a network of intersecting linear stress features on Mercury's surface, but these features have been poorly observed at best (Clark et al, 1988) possibly due to subsequent surface heating and erosional processes.

3.9 MERCURY'S HIGH BULK ABUNDANCE OF IRON

There are presently four major approaches advanced to explain the apparently high bulk abundance of iron not accounted for in the **equilibrium condensation** model. These invoke 1) Direct accretion of chemically reduced components of known composition (**Reduction/Oxidation** Model (Wanke, 1981) or **Enstatite Chondrite** Model (Wasson, 1988); 2) **Selective Accretion**: Differences in the response of iron and silicates to impact fragmentation and aerodynamic sorting in the presence of gas (Weidenschilling, 1978); 3) **Post-accretion Vaporization**: Preferential vaporization of silicates by solar radiation early in the Sun's evolution (Cameron 1985, Fegley and Cameron 1987); and 4) **Giant Impact**: Selective removal of silicate as a result of a Giant Impact on an already differentiated proto-planet (Wetherill, 1988, Benz *et al* 1988).

3.10 DIRECT ACCRETION OF REDUCED COMPONENTS

Ringwood (1966) and Wanke (1981) modeled Mercury's initial composition using two chemically distinct components, one highly reduced and moderately refractory, the other highly oxidized. Modeling based on chemistry when applied to the Earth and Mars, display a trend that predicts Mercury to be extremely reduced (Dreibus and Wanke, 1984, 1985). In the case of Mercury itself, the models generated are consistent with the implied reduced state of the planet's interior because of the absence of a surface FeO absorption band (since FeO would tend to fractionate with time in extruded melts).

Chondritic meteorites are the only materials known to have formed in the inner part of the Solar Nebula. Thus, they are likely to constitute samples of nebular material that accreted to form the terrestrial planets. Wasson (1988) suggests that the most reduced chondrites (EH and EL) formed nearest to the Sun while the most oxidized (CM and CI) formed furthest away. Based on existing knowledge of the chondrites in the inner Solar System, Wasson (1988) proposed that Mercury is composed of **enstatite chondrite.** No other process was thereby required to be present than that believed to have been operative during chondrite formation. However, the selective accretion of cores of differentiated asteroids is needed to produce an Fe/Si ratio 4 to 7 times greater than that observed in enstatite chondrites. Because Mercury's

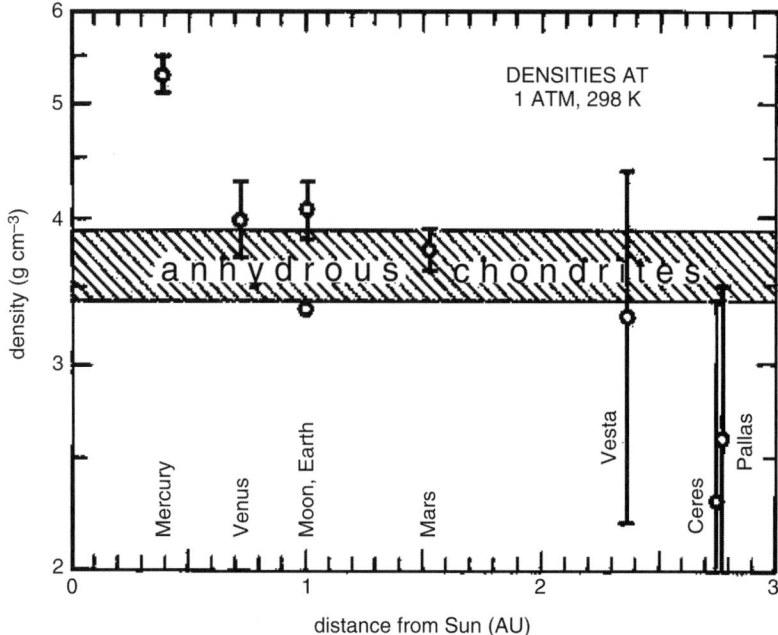

Figure 3-6. Plot of Density vs Distance from Sun for Terrestrial Planets showing that starting compositions are consistent with anhydrous chondrite except for Mercury which requires an enhancement in Fe by a factor of 4 (Wasson, 1988, The Building Stones of the Planets, in Mercury, Copyright 1988, The Arizona Board of Regents. Reprinted by permission of the University of Arizona Press.)

density is so high it requires enhancement in iron by a factor of 4 over the anhydrous chondritic composition consistent with other rocky bodies (**Figure 3.6**) (Wasson, 1988). The Wasson model contrasts strongly with the Lewis model, which calls for a high temperature in a narrow range between the condensation of Fe-Ni metal and Mg-rich olivine. The enstatite chondrite model joins many other models in predicting low FeO, but the enstatite chondrite model does not predict uniformly low volatile abundances.

3.11   THE SELECTIVE ACCRETION MODEL

Selective Accretion relies on differences between the aerodynamic and mechanical properties of metals in the inner Solar System (where the gas densities were relatively great and the time scale short) sufficient to support dynamic sorting leading to iron enrichment (Weidenschilling, 1978). Metal/silicate fractionation might have resulted from the **Selective Accretion** of the cores of differentiated asteroids. This process would require the original composition of the asteroids to be at least as reduced as

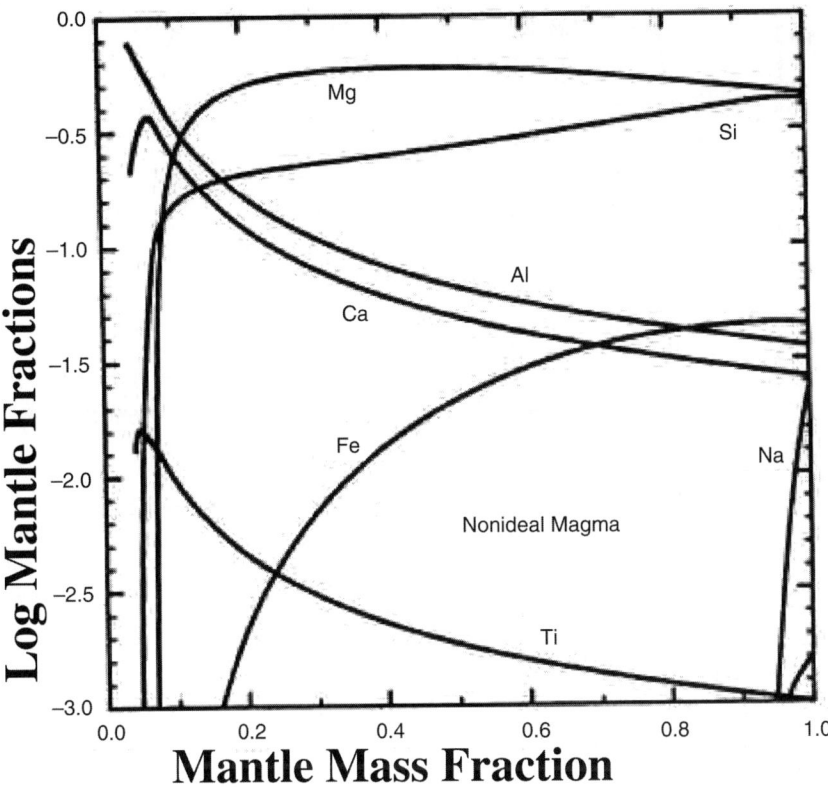

Figure 3-7. Change in Mantle Composition over Time based on vaporization of mantle and non-ideal magma (Cameron et al, 1988, The strange density of Mercury: Theoretical Considerations, in Mercury, Copyright 1988, The Arizona Board of Regents. Reprinted by permission of the University of Arizona Press.)

the H chondrites (the more reduced the composition the easier it would be to explain the implied Fe/Si fractionation). Alternatively, mechanical separation could have occurred in the solar nebula by precisely the same process (e.g. selective sticking, fragmentation, settling or radial drag) that produced the EH chondrites which have an Fe/Si ratio two times greater than the EL chondrites (Wasson, 1986).

3.12 POST-ACCRETION VAPORIZATION AND GIANT IMPACT MODELS

The **Post-accretion Vaporization** and **Giant Impact** hypotheses require removal of a large fraction of Mercury's silicate mantle either by the vaporization of silicates induced by T-Tauri phase of the young sun by

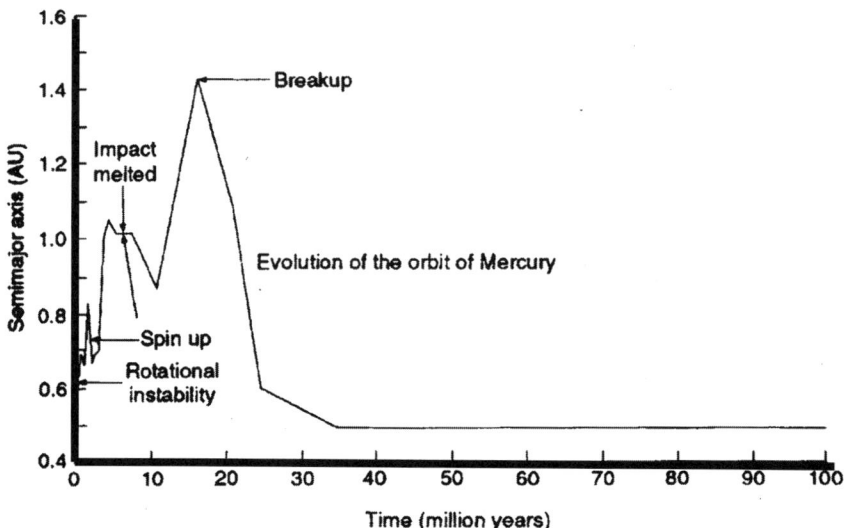

Figure 3-8. Impact of provenance (solar system position) on bulk composition of Mercury in context of other terrestrial planets during accretion: The variation in Mercury's semimajor axis during its formation is great and should thus have a great impact on its composition, including lower magnesium and iron silicate abundances, than if its position had remained closer to where it is today. (Wetherill, 1988, Accumulation of Mercury from planetesimals, in Mercury, Copyright 1988, The Arizona Board of Regents. Reprinted by permission of the University of Arizona Press.)

intensive, early bombardment (**Figure 3.7**) (Cameron, 1985; Cameron et al, 1988; Fegley and Cameron, 1987), or by ejection of the silicate mantle subsequent to collision with a planet-sized object (**Figure 3.8**) (Wetherill, 1987, 1988; Benz et al, 1987; Kaula, 1995; Vityazev et al, 1987). A greater fraction of the outer component is removed while the interior remains behind.

The **Giant Impact** Hypothesis requires the removal of a large fraction of Mercury's silicate mantle following collision with a planet-sized body while retaining the metallic core (Cameron, 1988). In another version of this model (Woolfson, 2000), Mercury is the heavily eroded satellite of a colliding protoplanet which was originally the mass of Mars. Ejected particles might have been reaccreted once they had cooled and passed through the vapor phase. However, it is more likely that they would have been of low enough density to be drawn into the Sun in a relatively short time, due to the Poynting-Robertson effect. The model would also lead to some depletion of the refractory oxides, a wide range of alkali oxide and volatile compositions, and an enhancement in FeO overall.

The fate of a particular body after a catastrophic impact event would, in the framework of such a model, depend on its size. Thus, larger, terrestrial-planet-sized, bodies like Venus and the Earth would continue to grow, while remaining close to the location where they were formed. Smaller bodies like Mercury and the Moon would, on the other hand, be subject to a wide migration of their semi-major axes during their growth phase (Wetherill, 1988). As a consequence, Mercury could accumulate materials that were originally formed over the entire Terrestrial Planet range of heliocentric distances. This process could have the effect of partially smoothing chemical differences associated with the primordial fractionation processes. However, the outcomes for particular bodies are not predictable and it was demonstrated by Lewis (1988) that simulations of the distribution of Solar System bodies lead to configurations that were substantially different from the one that is actually observed.

The **Impact Vaporization Model** predicts strong enrichment of refractories and the depletion of alkalis and FeO (Fegley and Cameron, 1987). According to the Impact Hypothesis, residual silicate material would be predominantly of mantle composition; the FeO content would reflect the oxidation state of the material from which the proto-planet accreted while the loss of much of the original crust would deplete Ca, Al, and the alkali metals without enriching the refractory elements.

According to both models, the present crust should primarily represent the integrated magma volume produced by partial melting of the relict mantle.

3.13   INFALL OF COMETARY/ASTEROID MATERIALS

Lewis (1988) considered what the consequence of a late in-fall of cometary and asteroid materials would have been on the individual models and concluded that, if the vaporization scenario is correct, the present surface inventories of FeO and alkali metals would be dominated by late in-falling asteroidal and cometary material. If the Giant Impact model is correct, then the surviving primordial FeO content is probably larger than the late in-fall contribution, (although in-fall could enhance the alkali abundances). In the case of preferential accretion, in-fall is probably not important as a source of either alkalis or FeO. Support for cometary in-fall as a mechanism comes from the unexpected indications that ice layers exist in permanently shadowed regions of the poles (Slade et al, 1992).

3.14   DISCRIMINATION BETWEEN THE MODELS

All models predict at least some depletion of volatiles (See **Figure 3.3**). No matter what the model, low temperature condensates would be higher near the surface, so volatiles would always be more easily swept away (Lewis and Prinn, 1984). The **Post-accretion Vaporization** model leads to severe depletion of alkali oxides, volatiles, and FeO and extreme enrichment of refractory oxides. The **Giant Impact** model would lead to some depletion of refractory oxides, a range of alkali oxide and volatile compositions, and enhancement of FeO. In the **Giant Impact** model, accretion takes place over a much larger portion of the inner solar system, with final composition being most greatly influenced by the range and mean of the semi-major axis of the accreting body (**Figure 3.8**) (Strom, 1997; Wetherill, 1994). Thus, although the outcomes are less systematic and predictable, the sulfur depletion problem could be less severe. On the other hand (Lewis, 1988), simulations of the **Giant Impact** model produce planetary systems with far less massive and more numerous bodies than our own; in addition, the large excursions these bodies take in their lifetimes might wipe out any differences like the ones we observe in our solar system. Taylor (2001) raised the issue that vaporization of the silicate mantle is not consistent with an alkali atmosphere and that collisional removal is more likely.

At present, Mercury's bulk composition is largely unconstrained by observations, and it is envisioned in the literature in a range that goes from extremely refractory to volatile-rich (Goettel, 1988). Ground based reflectance spectra at visible, near infra-red and millimeter wavelengths suggest low FeO (~3%) and alkali bearing feldspar contents (Vilas, 1988; Sprague et al, 1994, 1997, 2002; Jeanloz et al, 1995; Blewett et al, 1997). Na and K measurements in Mercury's exosphere suggest significant alkali content but it is not known if these species come from a veneer of meteoritic material or from deeper regions of the crust (Hunten et al, 1988).

The assumption that Mercury is enriched by iron is based on indirect evidence. Processes governing the composition of the interior may be largely de-coupled from processes that determine the core/mantle ratio. In this regard, the **Giant Impact** and the **Selective Accretion** processes would change the core/mantle ratio in a way that is independent of the starting composition.

No unique composition leads to the derived mean density (**Table 3.3**) (Goettel, 1988). For many years, the refractory-rich **Equilibrium Condensation** scenario was favored. All except the **Selective Accretion** model predict a moderately refractory-rich Mercury. A volatile-rich model would assume that the silicate portion of Mercury has the same composition as the SNC meteorite parent body (e.g., Mars). Sulfur would be available to form a lower melting point alloy in a large core but the abundance of FeO would also be high in the mantle in this case, which is a disadvantage, although it could be supposed to subsequently be depleted in a volatilization episode. Goettel (1988) suggested that some combination of the models is required in order to avoid the un-resolvable problems that extreme compositions present.

Fortunately, each model implies distinctive mechanisms for the evolution of the crust, from which surface composition can be inferred (**Table 3.3**).

Thus, each scenario offers an hypothesis that is, in principle, ultimately testable on the basis of compositional measurements made using orbiting spectrometers (Lewis, 1988).

Table 3-3. **Model Predictions for Refractory to Volatile-Rich Mercury (Goettel, 1988)**

Constituent		Refractory-Rich	Preferred Now	Volatile-Rich
Mantle	Al2O3	16.6	3.5-7	3.3
	CaO	15.2	3.5-7	3.0
	MgO	34.6	32-38	32
	SiO2	32.3	38-48	45
	FeO	0	0.5-5	15
	Na2O	0	0.2-1	1.4
	H2O	0	A little	A lot
Core	Fe	92.5	88-91	76
	Ni	7.5	6.5-7.5	6.2
	S	0	0.5-5	17.6

Selective Accretion Model: Mercury's silicate portion should contain a few percent Al.; about 1% alkali oxides (Na and K); and could be quite variable in FeO. Sulfur should be nearly absent unless the Solar Nebula temperature in the region of formation were lower than is predicted by the Equilibrium Condensation Model (which would also enhance the FeO content). If Mercury's magnetic field is a remnant field and the core is completely solidified, a very low abundance of sulfur would be acceptable.

Post Accretion Vaporization: Alkali oxides would be virtually non-existent and FeO very low. Refractory oxides would be high and easy to measure.

Giant Impact: Alkali and refractory oxides would be low and highly variable, with FeO present in low to moderate abundances.

Determination of the bulk chemistry of the silicate portion of Mercury can provide an opportunity to gain an insight as to which of the mechanisms operating during the formation of the inner Solar System had the greatest influence on the bulk composition of the inner planets.

Potentially, atmospheric abundances of refractories and volatiles in the atmosphere could provide constraints on bulk composition and thus on models for Mercury's origin. Atmospheric abundances are the result of interactions of high energy particles with the surface in ways which are complex and not necessarily stoichiometric. This will be discussed in more detail in Chapter 5 on Mercury's Atmosphere.

3.15   SUMMARY

No direct compositional measurements are available for Mercury, thus little is directly known about its interior processes. However, several discoveries made by Mariner 10 lead to certain inferences on its interior composition and structure: The planet has high density, indicating the existence of an iron-rich core, as well as a magnetic dipole, indicating the presence of an active dynamo.   These two findings point to extensive geochemical differentiation during core formation and intensive tectono-volcanic activity requiring and resulting from the existence of a partially molten interior. Several proposed models of solar system formation are consistent with the planetary compositional data obtained to date (for the Earth, Mars, asteroids, the outer planets), but predict a wide variety of compositions for innermost planet Mercury. Solar system origin will remain poorly understood until we obtain direct information on Mercury's interior.

3.16   REFERENCES

Anderson, J. D., G. Colombo, P. B. Esposito, E. L. Lau, and G. B. Trager, The mass, gravity field, and ephemeris of Mercury, *Icarus*, **71**, 337-349, 1987.

Anderson, J. D., R. F. Jurgens, E. L. Lau, M. A. Slade III, and G. Schubert, Shape and orientation of Mercury from radar ranging data, *Icarus*, **124**, 690- 697, 1996.

Benz, W., W.L. Slattery and A.G.W. Cameron, Collisional stripping of Mercury's Mantle, *Icarus* **74**, 516-528, 1988.

Blewett, D.T., P. Lucey, B.R. Hawke, G. G. Ling, M.S. Robinson, A comparison of Mercurian reflectance and spectral quantities with those of the Moon, *Icarus*, **129**, 217 -231, 1997.

Branham, The mass of Mercury, *Planet. Space Sci*, **42**, 213-219, 1994.

Burns, J.A., 1976, Consequences of the tidal slowing of Mercury, *Icarus*, 28, 453-458.

Cameron, A.G.W , The partial volatilization of Mercury, *Icarus* **64**, 285-294, 1985.

Cameron, A., B. Fegley, W. Benz, and W. Slattery, The strange density of Mercury: Theoretical considerations. In *Mercury, Eds.* Vilas, Chapman, and Matthews, (Publ, U. Arizona) pp. 692-708, 1988.

Campbell, W.H., *Introduction to Geomagnetic Fields* (Publ. Cambridge University Press), 291 p., 1997.

Cassen, P., R. E. Young, G. Schubert, and R. T. Reynolds, Implications of an internal dynamo for the thermal history of Mercury, *Icarus*, **28**, 501-508, 1976.

Christon, S., A comparison of the Mercury and Earth magnetospheres: Electron measurements and substorm time scales, *Icarus*, **71**, 448-471, 1987.

Clark, P. and J. Trombka, Remote X-ray spectrometry for NEAR and future missions: Modeling and analyzing X-ray production from source to surface, *J. Geophys. Res., R*, 16361-16384, 1997.

Clark, P., M. Leake, R. Jurgens, Goldstone radar observations of Mercury. In *Mercury*, Vilas, Chapman, Matthews, Eds., U. Arizona Press, 77-100, 1988.

Connerney, J. E. P., and N. F. Ness, Mercury's Magnetic Field and Interior. In *Mercury*, Eds. Vilas, Chapman, and Matthews (Publ. Univ. Arizona Press), pp. 494-513, 1988.

Connerney, J. E. P., M.H. Acuna., P.J. Wasilewski, G. Kletetschkan, N.F. Ness, H. Reme, R.P. Lin and D.L. Mitchell, The Global Magnetic Field of Mars and Implications for Crustal Evolution, *GRL,* **28,** Issue 21, 4015-4018, 2001.

Cook, A.H., *Interiors of the Planets* (Publ. Cambridge University Press), 348 p., 1982.

Dreibus G. and H. Wanke, Accretion of the Earth and the inner planets; *Proc. 27th Intern. Geol. Con.*, Vol. **11,** *VNU Science Press*, 1-20, 1984.

Dreibus G. and H. Wanke, Mars, a volatile-rich planet; *Meteoritics* **20,** 367-381. 1985.

Fegley, B. Jr and A.G.W. Cameron, A vaporization model for iron/silicate fractionation in the Mercury protoplanet, *Earth Planet. Sci. Lett.* **82**, 207-222, 1987.

Fricker, P. E., R. T. Reynolds, A. L. Summers, and P. M. Cassen, Does Mercury have a molten core? *Nature* **259,** 293-294, 1976.

Goettel, K. Present bounds on the bulk composition of Mercury: Implications for planetary formation processes. In *Mercury*, Eds. Vilas, Chapman, and Matthews (Publ. Univ. of Arizona), pp. 613-621, 1988.

Gubbins, D., Speculations on the Origin of the Magnetic Field of Mercury, *Icarus*, 30, 186-191, 1977

Hagermann, A., Planetary Heat Flow Measurements, *Philosophical Transactions of the Royal Society A: Mathematical, Physical and Engineering Sciences*, Volume 363, Number 1837, 2777-2791, 2005.

Hauck, S.A., Dombard, A.J., Phillips, R.J., Solomon, S.C., Internal and tectonic evolution of Mercury, *EPSL*, 222, 713-728, 2004.

Hunten, D.M., Morgan, T.H. and Shemansky, D.H., The Mercury Atmosphere in *Mercury*, Eds. Vilas, Faith, Chapman, Clark R. and Shapley Matthews, Mildred (Publ. Univ. of Arizona Press, Tucson) pp. 562–612, 1988.

Jeanloz, R., and D. Mitchell, A. Sprague, I. De Pater Evidence for a basalt-free surface on Mercury and implications for internal heat. *Science*, **268**, 1455-1457, 1995.

Kaula, W.M., Formation of the Terrestrial Planets, *Earth Moon and Planets*, 67, 1-3, 1-11, 1995.

Keil K., Mineralogical and chemical relationship among enstatite chondrites; *J. Geophys. Res.* **73**, 6945-6976. 1968.

Lewis, J.S., Metal/silicate fractionation in the solar system *Earth Plan Sci Lett,* **15,** 286- 292, 1972.

Lewis, J. S., Origin and composition of Mercury. In *Mercury*, Eds. Vilas, Chapman, and Matthews (Publ. Univ. Arizona Press), pp. 651-666, 1988.

Lewis, J.S., *Physics and Chemistry of the Solar System*, 2nd Ed. (Publ. Elsevier), 655p., 2004.

Lewis, J.S. and R.G. Prinn, *Planets and their Atmospheres Origin and Evolution* (Publ. Academic Press), 470p., 1984.

Lyttleton, R.A., History of the mass of Mercury, *QJR Astro Soc*, **21**, 400-413, 1980.

Lyttleton, R.A., More thoughts about Mercury, *QJR Astro Soc*, **22**, 322-323, 1981.

Margot, J., Peale, S.J., Jurgens, R.F., Slade, M.A., Holin, I.V., 2004, Earth-based Measurements of Planetary Rotational States, *EOS Trans*, AGU, 85 (47), Fall Meet. Suppl., Abstract G33A-02, 2004.

Melosh, J. and W. McKinnon, The tectonics of Mercury. In *Mercury*, Eds. F. Vilas, C. Chapman, and M. Matthews (Publ. Univ. Arizona Press), pp. 374-400, 1988.

Ness, N., K. Behannon, R. Lepping, Y. Whang, and K. Schatten, Magnetic field observations near Mercury: Preliminary results from Mariner 10, *Science*, **183**, 1301 -1306, 1974.

Ness, N., K. Behannon, R. Lepping, Y. Whang, and K. Schatten, The magnetic field of Mercury, *J. Geophys. Res.* **80**, 2708 – 2716, 1975.

Peale, S. J., Does Mercury have a molten core? *Nature*, **262**, 765-766, 1976.

Peale, S. J., Rotational dynamics of Mercury, in *Mercury,* Eds. Vilas, Chapman and Matthews (Publ. Univ Arizona Press), pp. 461-493, 1988.

Peale, S.J., Free Rotational Motions of Mercury, *EOS Trans*, AGU, 85 (47), Fall Meet. Suppl., Abstract G33A-03, 2004.

Ringwood, A., Chemical evolution of terrestrial planets, *Cosmochim Acta* **30**, 441-504, 1966.

Russell, C.T., D.N. Baker, and J.A. Slavin, The Magnetosphere of Mercury. In *Mercury* Eds. Vilas, Chapman and Matthews (Publ. Univ. of Arizona Press, Tucson) pp. 514-561, 1988.

Schubert, G., M. N. Ross, D. J. Stevenson, and T. Spohn, Mercury's thermal history and the generation of its magnetic field. In *Mercury*, Eds. Vilas, Chapman, and Matthews (Publ. Univ. Arizona Press) pp. 429-460, 1988.

Sigfried, R. W. and S. C. Solomon, Mercury: Internal structure and thermal evolution, *Icarus*, **23**, 192-205,1974.

Slade, M. B. Butler, D. Muhleman Mercury radar imaging: Evidence of polar ice. *Science* 258, 635-640. 1992.

Spohn, T., F. Sohl, K. Wieczerkwoski, V. Conzelmann, The interior structure of Mercury: what we know, what we expect from Bepi Colombo, *Planet. Space Sci,* **49**, 1561-1570, 2001.

Sprague, A.L., R. Kozlowski, F. Witteborn, D. Cruikshank, D. Wooden, Mercury evidence for anorthosite and basalt from mid infrared (7.3 – 13.5 micron) spectroscopy, *Icarus*, 109, 156-167, 1994.

Sprague, A.L., D. Nash, F. Witteborn, D. Cruikshank, Mercury's feldspar connection mid-IR measurements suggest plagioclase, *Adv in Space Research*, 19, 1507-1510, 1997.

Sprague, A.L., J.P. Emery, K.L. Donaldson, R.W. Russell, D.K. Lynch, A.L. Mazuk, Mercury: Mid-infrared (3-13.5um) observations show heterogeneous composition, presence of intermediate and basic soil types, and pyroxene, *Met Plan Sci,* **37**, 1255 -1268, 2002.

Stevenson, D., Dynamo generation in Mercury, *Nature,* **256**, 634, 1975.

Stevenson, D., Applications of liquid state physics to the Earth's core. *Phys. Earth Planet. Int.,* **22,** 42-52, 1980.

Stevenson, D., Mercury's magnetic field: a thermoelectric dynamo? *Earth Planet. Sci. Lett.,* **82,** 114-120, 1987.

Strom, Robert G. and Ann L. Sprague in *Exploring Mercury, the Iron Planet* (Publ. Springer in association with Praxis Publishing), 2003.

Taylor, S.R., The chemical evolution of the galaxy, *Nature,* **414**, 6861, 253, 2001.

Van Flandern, T. and R. Harrington, Dynamical investigation of conjecture that Mercury is an escaped satellite of Venus, *Icarus*, 28, 435-440, 1976.

Vilas, F. Surface composition of Mercury from reflectance spectrophotometry. In *Mercury*, Eds. Vilas, Chapman, Matthews (Univ. Arizona Press), pp. 59-76, 1988.

Vityazev , A.V., G.V. Pechernkova and V.S. Safronov, Formation of Mercury and removal of its silicate shell, In *Mercury*, Eds. Vilas, Chapmen, Matthews (Univ. Arizona Press), pp. 667-669, 1988.

Wanke H., Constitution of terrestrial planets; *Phil. Trans. R. Soc.Lond.* A **303**, 287-302, 1981.

Ward, W., G. Colombo, F. Franklin, Secular resonance, solar spin down, and orbit of Mercury, *Icarus*, 28, 441-452, 1976.

Wasson, J. T., The building stones of the planets. In *Mercury*, Eds. Vilas, Chapman, Matthews (Univ. Arizona Press), pp. 622-650, 1988.

Weidenschilling S.J., Iron/silicate fractionation and the origin of Mercury, *Icarus*, **35**, 99-111, 1978.

Wetherill, G., Accumulation of Mercury from planetesimals in *Mercury*, Eds. F. Vilas, Chapman, and Matthews (U. Arizona Press) 670-691, 1988.

Wetherill, G., Provenance of the Terrestrial Planets, *GeoChim et Cosmochim Acta,* **58**, 4513-4520, 1994.

Woolfson, M.M., *The Origin and Evolution of the Solar System* (Publ. The Institute of Physics, London), 420 p., 2000.

1.17 SOME QUESTIONS FOR DISCUSSION

1. What observational evidence could provide a basis for confirming a model for the formation of Mercury?

2. How could Mercury have the density it does as well as an apparent magnetic dipole field and not have a dynamo?

3. Why does the equilibrium condensation hypothesis alone not explain observed physical properties of Mercury?

4. What range of bulk compositions would explain the observations of Mercury right now?

5. Consider the implications for C_m/C for each formation model.

Chapter 4

MERCURY'S SURFACE

4.1 PRESENT UNDERSTANDING OF MERCURY'S SURFACE

Current models of the nature and evolution of Mercury's surface are not well constrained. Our understanding of processes which shaped the character of Mercury's surface, including its physical and compositional nature, were, until recently, almost exclusively based on analyses of Mariner 10 data (**Figures 4-1a and 4.-b**) (Strom and Sprague, 2003). Included in the Mariner 10 dataset are images covering most of one hemisphere with resolution no better than 100 meters and with minimal compositional information indicating low iron abundance and relatively little difference between major terranes (e.g., Hapke et al, 1994; Robinson et al, 2000; Strom and Sprague, 2003).

Ground-based observations at thermal, microwave, and radio wavelengths have confirmed the presence of a lunar-like regolith (Ledlow et al, 1992; Mitchell and DePater, 1994) observed during the Mariner 10 mission (Chase et al, 1976). Although the original compositional character has been confirmed by ground-based spectral reflectance measurements (Vilas et al, 1984; Vilas, 1985), some workers have interpreted recent ground-based mid-IR spectra to indicate the presence of lunar breccia or enstatite achondrite like materials (Tyler et al, 1988; Sprague et al, 1994). Most strikingly, ground-based radar reflectivity maps, equatorial topography, and roughness profiles (**Figures 4-2 and 4.3**) suggest regional heterogeneity and hemispherical asymmetry (Clark et al, 1988; Harmon and Slade, 1992), at least in the equatorial region, which has recently been confirmed by Anderson and coworkers (1996), basin-like features in the unimaged hemisphere (Goldstein, 1970, 1971; Zohar and Goldstein, 1974; Harmon,

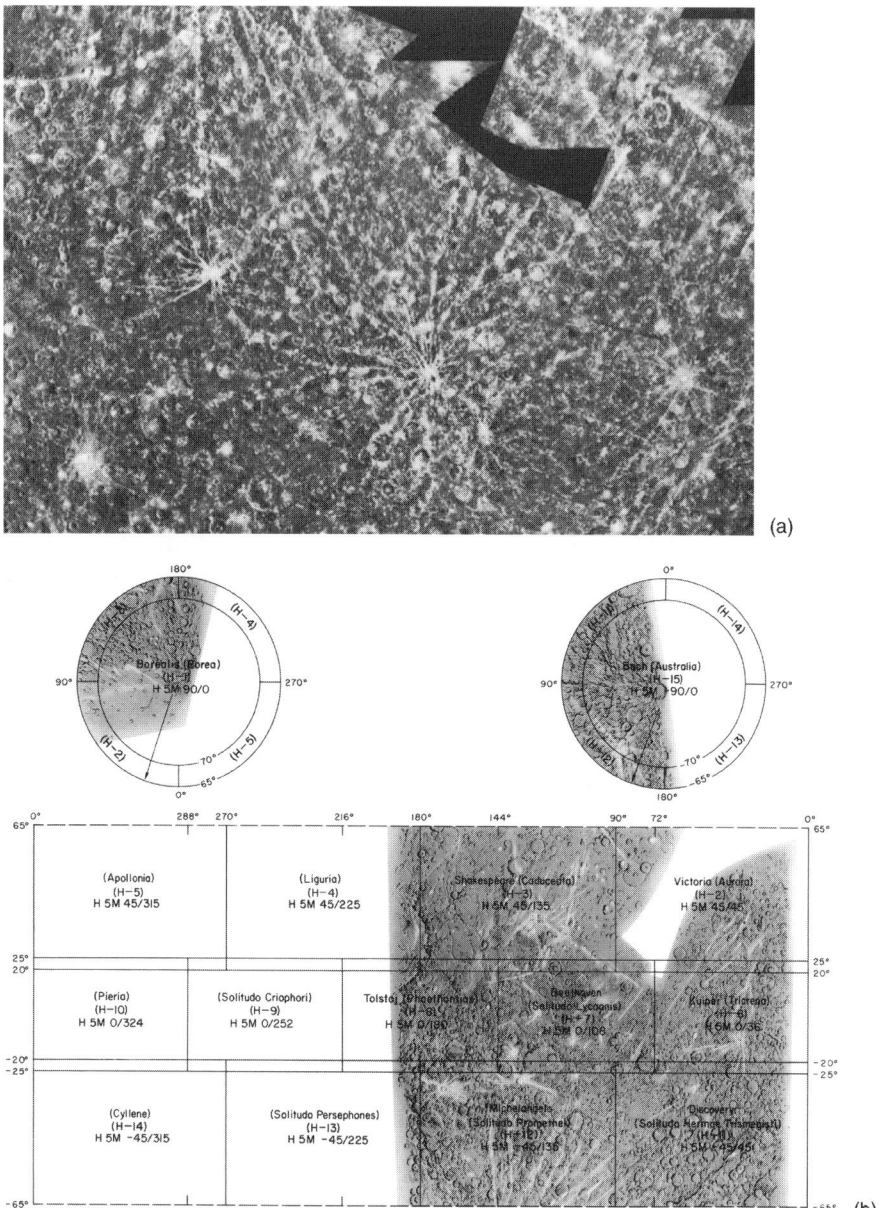

Figure 4-1. (a) Mariner 10 H7-Cenetered Photomosaic from the NASA Atlas of Mercury SP432. Particularly noteworthy are the orthogonal bright lineaments, the most striking albedo f eature, in some but not all cases tied in with bright crater rays. Lacking are clear indications of dark plains in this hemisphere so prominent on our Moon. (b) USGS Shaded Relief Geological Map of Mercury indicating that less than 50% of the surface, the central portion of which is seen in (a) above, is imaged at resolutions adequate enough to discern typical features and terranes.

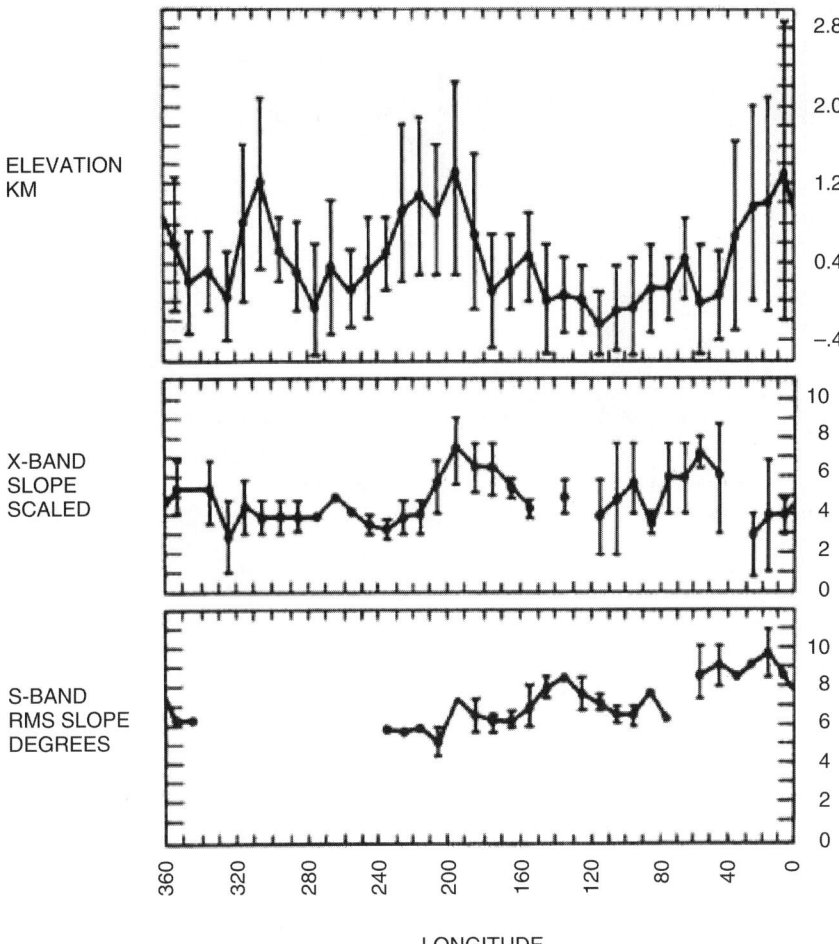

Figure 4-2. Comparison of topography and scattering properties at the centimeters (X-band) and tens of centimeters scales for 10 degree longitude bins in the equatorial region, with errorbars equivalent to standard deviation. (Clark et al, 1988, Goldstone Radar Observations of Mercury, in Mercury, Copyright 1988, The Arizona Board of Regents. Reprinted by permission of the University of Arizona Press.)

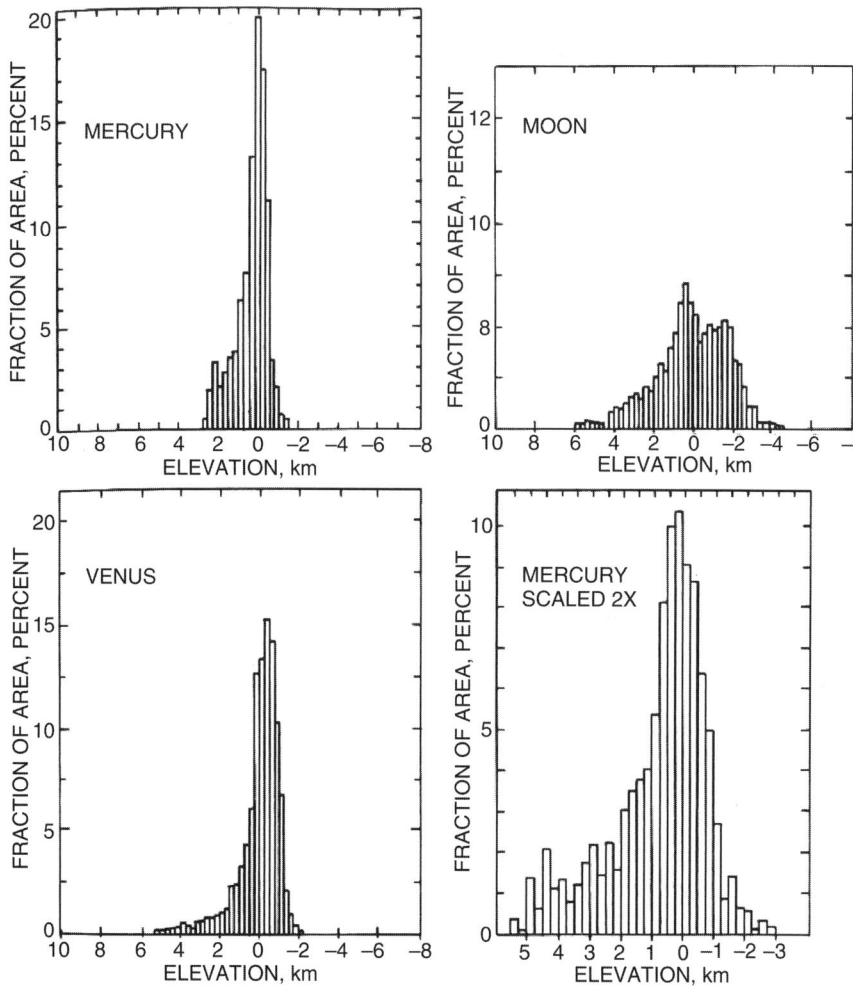

Figure 4-3. Comparison of Equatorial Region Topographies of Mercury, the Moon, and Venus. Note rough similarity between Mercury and Venus. When Mercury topography is scaled by a factor of two for comparison to the Moon to account for differences in gravity, Mercury has a distinctively lunar highland like appearance. (Clark et al, 1988, Goldstone Radar Observations of Mercury, in Mercury, Copyright 1988, The Arizona Board of Regents. Reprinted by permission of the University of Arizona Press).

1997), and polar deposits (Slade et al, 1992; Harmon and Slade, 1992; Harmon et al, 2001). On the other hand, little evidence for asymmetry in the distribution of large-scale albedo features can be seen in the most recent ground-based low resolution imaging of Mercury (Warell and Limaye, 2001).

4.2 PHYSICAL PROPERTIES OF THE SURFACE AND REGOLITH

At first glance, Mercury and the Moon appear quite similar in appearance. Both are covered with impact generated regolith, consisting of a centimeters thick insulating and diurnally variable layer on top of a meters thick compacted layer with higher thermal conductivity. Thus, their very similar optical properties are not surprising. The range of albedos of the two bodies is overlapping, but Mercury tends to have somewhat higher and more uniform albedo. The absence of the opaque darkening agent ilmenite may account for the overall higher albedo of the planet's surface (Taylor, 2001).

What are the differences between Mercury and the Moon, as indicated in **Table 4-1**, and are they significant (Strom and Sprague, 2003)? Evidence comes from a number of sources including observations, largely ground-based, in the visible, infrared, microwave, and radio spectrum (Vilas, 1985, 1988; Chase et al, 1976; Ledlow et al, 1992; Mitchell and DePater, 1994; Warell and Limaye, 2001; Warell, 2001, 2002) as well as simulations of impact-induced effects (Cintala, 1992).

Table 4-1. **Optical Properties of Mercury and the Moon**

Planet	Albedo at 554 nm 5 degree phase			Maturity slope from 650-950 nm	Backscatter
	Lunar Maria or Mercury Smooth Plains	Lunar Highlands or Mercury Intercrater Plains	Young craters		
Mercury	.12-.13	.16-.18	.36-.41	Higher, more altered	Higher at all phases
The Moon	.06-.07	.10-.11	.15-.16	Lower, more altered	Lower at all phases

The typically higher slopes and redness characterizing spectral reflectance plots, suggest a greater maturity of the Mercury than of the lunar regolith (Vilas, 1988). Also, Mercury displays higher back-scatter at all phase angles. These data are interpreted to indicate the presence on the planet of smaller grains characterized by high transparency (few are opaque), and complex faceting that is consistent with a high agglutinate content (Vilas, 1985, 1988). The degree of linear polarization at visible wavelengths indicates the presence of a fine particulate layer, or regolith, at the surface and appears to have a longitude dependence on Mercury, indicating physical or even mineralogical differences in this regolith (Meierhenrich et al, 2002). Dynamic models for regolith production (Mendell and McKay, 1975; Cintala, 1992) as well as polarimetric observations (Leake and Dollfus, 1986) indicate that a grain sizes are smaller on Mercury than the Moon. Grains range from 10 to 80 microns and a greater proportion of soil grains are in the smallest size range (Leake and Dollfus, 1986).

Thermal, microwave and radio observations of Mercury are primary sources of information on the regolith because thermal emissions resulting from physical properties of the regolith dominate at these wavelengths. Mariner 10 thermal infrared observations of the nightside provided an early basis for understanding the nature of Mercury's regolith (Chase et al, 1976). During the long night, such thermal variations are dominated by regolith thermal properties. Based on these data, Mercury has a regolith with lunarlike properties (Mitchell and de Pater, 1994; Ledlow et al, 1992). The derived dielectric constant and loss tangent are both consistent with a low density particulate layer, and a gradual increase in density with depth on a centimeter scale (Mitchel and de Pater, 1994; Ledlow et al, 1992). The loss tangent and microwave opacity are somewhat lower than for the Moon, probably due to lower Mercurian abundance of mafic minerals. Also, the flatness in the emission spectrum from centimeter to tens of centimeter range may be a measure of the relative transparency of the regolith. The infrared emissivities have been shown to be not too different from, although somewhat lower than, those of the lunar regolith. Radar-derived centimeter to tens of centimeter scale RMS (Root Mean Square) slopes are indicative of lunarlike surface roughness on Mercury (Pettengill, 1987). Mercury's less than lunar polarization ratio, which is comparable to that of Mars or Venus, is thought to be due to the smaller fraction of the surface covered with fresh, dense ejecta blankets (Harmon, 1997).

Profiles of thermal properties **(Figure 4-4)** indicate that the thermal inertia increases 1.5 times from 180 to 360 degrees longitude across the unimaged hemisphere, and that the soil density or rock abundances increases by a factor of two at longitudes 236, 247, 264, and 290 degrees. Interestingly, all of these locations have been demonstrated to have high radar reflectivity indicating higher populations of rocks on the scale of tens of centimeters where ground-based radar reflectivity and polarization data are available (Harmon, 1997). Radar-derived roughness parameters have been shown to be greater in the unimaged hemisphere (Clark et al, 1988). Radio anomalies at the 'hot poles' may indicate the presence of complex terrane, which would correlate with the presence of Caloris and the antipodal terranes (Ledlow et al, 1992; Mitchell and dePater, 1994)).

Simulations of impact-induced thermal processes, including melting and vaporization, have indicated that on Mercury far more material is melted and

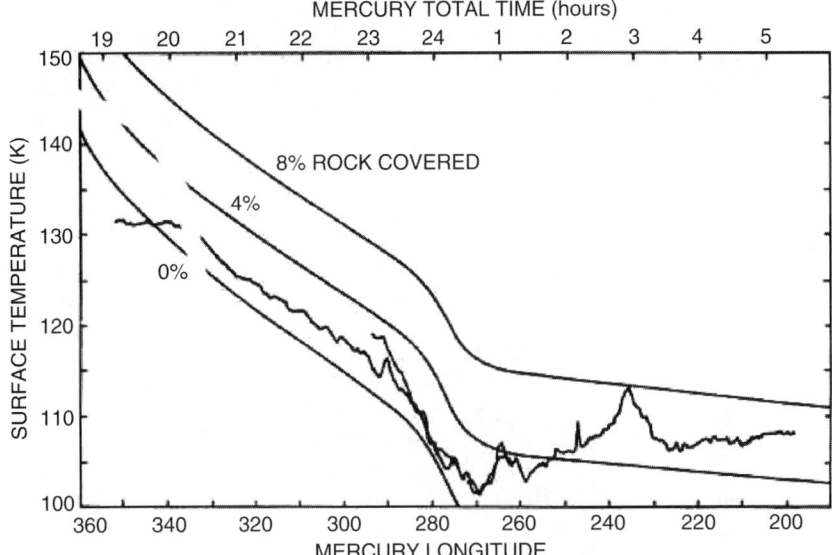

Figure 4-4. Mariner 10 Thermal IR Profile of nighttime surface temperatures and derived rock cover, for surface of thermal inertia 0.0016 and rock of thermal inertia 0.05 cal cm^{-2} sec$^{-1/2}$ K^{-1} (Reprinted from Chase et al, 1976, with permission from Elsevier).

vaporized at impact due to greater velocity and frequency of impact events (greater gravity) as well as the far greater temperature of Mercury's surface (Cintala, 1992). As a result, Mercury's regolith should have a much greater agglutinate content, and experience far higher gardening (overturn) rate, lower solar wind exposure, H implantation, and potential reduction of iron, more rapid 'maturation' and 'blanding' away of differences. All of these trends, when combined, should result in less intrinsic darkening and albedo variation observed.

Recent observations made by workers of the Uppsala Observatory (Warell and Limaye, 2001; Warell, 2001, 2002) reveal albedo and color information in the form of a global Minnaert normalized map with a resolution of approximately 200 km for the entire surface of Mercury, including the hemisphere that remained unimaged by Mariner 10 (**Figure 4-5**). Uppsala and Mariner 10 data are in good agreement in areas of overlapping coverage. At this resolution, no regional and hemispheric scale asymmetries were observed in the distribution of bright albedo features, most of which are recent impact craters. However, large structures identified by ground-based radar observations have been confirmed in the Uppsala Observatory observations.

Recent ground-based optical imaging done by other workers at Skinakas Observatory (Ksanfomality et al, 2003) are in agreement with Mariner 10 observations where overlapping coverage exists, and, even more exciting,

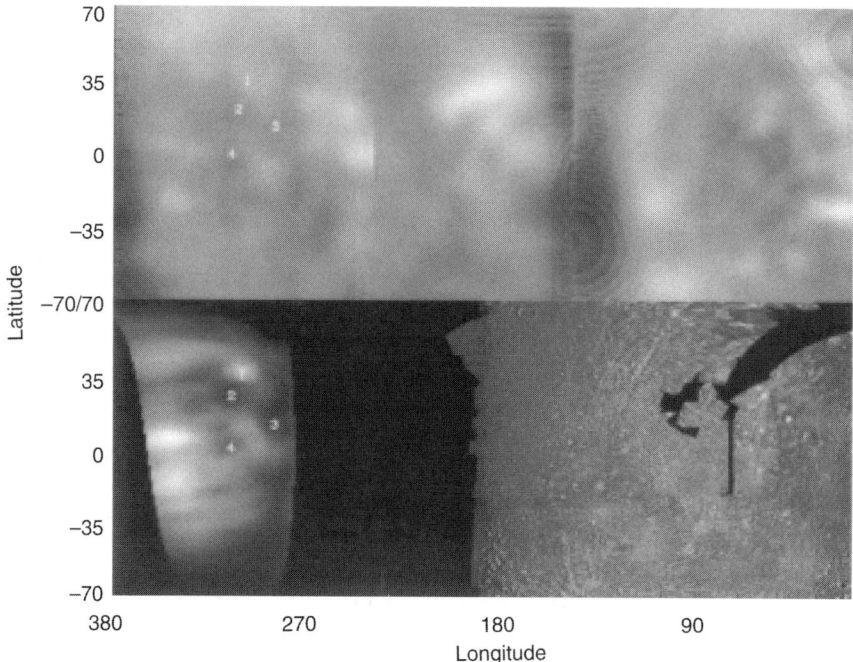

Figure 4-5. Ground-based Minnaert images of Mercury with labeling of prominent unimaged hemisphere albedo features.   Ground-based data are compared with the Mariner 10 photomosaic. Reprinted from Mendillo et al, 2001, with permission from Elsevier.

reveal a multi-ring feature centered at approximately 270 degrees longitude in the hemisphere unimaged by Mariner 10 **(Figure 4-6)**. These images have 100 km resolution.

4.3   COMPOSITION OF MERCURY'S SURFACE AND REGOLITH

As we discussed in the section on the interior, surface compositional data from Mercury is sparse. Assessments of mineralogy were originally based on color differences established with broadband color filters (Hapke et al, 1975, 1994). Recent reanalysis of these data (Robinson et al, 1992, 1997, 2000) have led to great improvements in signal to noise ratio and sensitivity of the two useful parameters, the ratio of UV to orange color band and orange albedo **(Figure 4-7)**. Robinson and coworkers (1997) removed artifacts from, resampled each pixel at the sub-spacecraft point, and renormalized to the latest Hapke functions (Hapke, 1986). They then interpreted the improved results by assuming a regolith with lunarlike properties, where the degree of darkening is related to exposure and

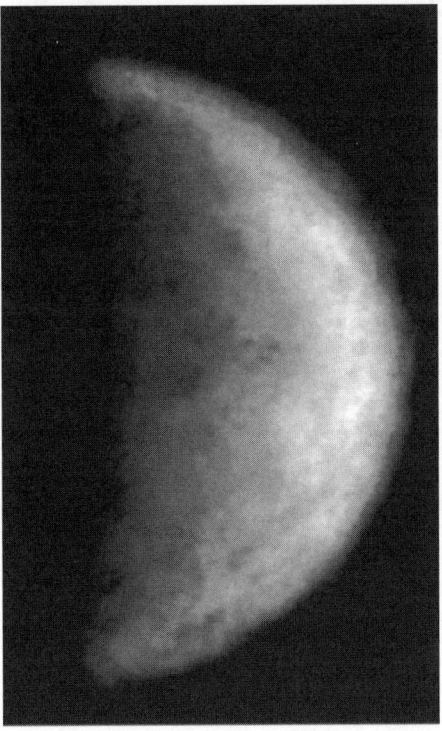

Figure 4-6. Ground-based Optical Image of Portion Unseen by Mariner 10 (210 to 285 degrees longitude) taken in May of 2002 by Ksalfomality and coworkers (2003). Note multi-ring basin named 'Skinakas' after the observatory in Greece where the images were taken. Resolution is approximately 100 km. Reprinted with permission of Kluwer Academic Press.

variations in the degree of reddening to titanium, or ilmenite, abundance. In this case, the color band ratio changes due to slope steepening as an iron silicate bearing soil matures. In the image of this band, crater rays, which would be more recent and less mature, are prominent. On the orange albedo image, basins show varying degrees of brightness, possibly indicative of compositional variations of the exposed strata. Some unambiguous color boundaries associated with geomorphological boundaries of the plains and geological units associated with basin complexes, which would be diagnostic of volcanic origin for the plains, are observed. The nature of Mercury's silicate crust was confirmed as being heterogeneous on the scale of Mariner 10 observations.

In particular, a region with deposits of apparently different rock compositions has been identified. Some of these deposits, known as LBOs, are low in albedo and opacity and blue in color. LBOs are found in the vicinity of Rudaki in the Tolstoj plains, Homer Basin, and Lermontov Basins. Other deposits, high in orange albedo, are found in the Kuiper Muraski complex (Robinson and Hawke, 2001; Robinson et al, 2001).

Figure 4-7. Recalibrated Mariner 10 color composite (Robinson and Lucey, 1997) of incoming hemisphere showing distinct correlation between color and geomorphological boundaries for Mercurian plains: Red is inversely correlated with opaque mineral content, green with maturity, and blue with visible color. Rudaki Crater, the large diffuse orange area near the lower left hand corner, contains plains of distinctly different color than the surroundings which apparently embay portions of the boundary, indicating emplacement as a fluid, or lava. Distinct bright orange areas have the characteristics of pyroclastic deposits. (From MESSENGER Web site courtesy of APL.) **See Color Plate 1.**

LBOs are interpreted to be pyroclastic deposits, indicating volcanic origin, based on the patchy nature of their deposition. Kuiper Muraski is thought to be similar to lunar anorthositic crust.

Ground-based observations in the visible and near infrared show a similarity to mature lunar soils and a weak to non-existent spectral feature associated with iron-bearing minerals, limiting the abundance of the oxidized iron to less than a couple of percent (**Figure 4-8**) (Vilas, 1985; Robinson and Taylor, 1984; Warell, 2002), similarly to the lunar highlands. Gaffey and McCord (1978) have suggested that Mercury's surface is dominated by the iron-free silicate, enstatite. Iron could still be present as free iron of course with little impact on reflectance spectra.

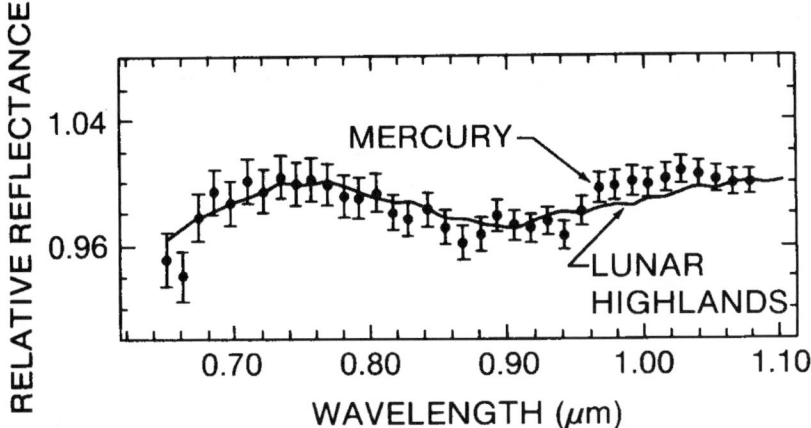

Figure 4-8. Near IR reflectance spectra of Mercury, covering intercrater and smooth plains terrane, and the lunar highlands soil (McCord and Clark, 1979, Copyright AGU) showing indication of the 0.9 iron band for orthopyroxene. Reproduced by permission of AGU.

Warell (2001, 2002) observed a steep spectral slope between 650 and 940 nm which should be comparable to that of mature lunar anorthositic soils, although no such measurements have been made to date. Much lower slopes have been derived from the immature lunar anorthositic soils (Blewett et al, 1997). The closest lunar match, in terms of spectral slope and its relationship to emission angle, is the mature Apollo 14 (Fra Mauro) soil. Mercury shows greater uniformity in albedo, color variation, and scatter, as well greater backscatter than the Moon in this spectral region (Warell, 2001). Very modest differences between Mercury spectra obtained at different sub-Earth longitudes have raised the possibility of local variations, but no consistent differences in composition between terranes or hemispheres have been observed in this spectral region (Vilas and McCord, 1976; Vilas et al, 1984; Vilas, 1985; Warell, 2001). These observations are consistent with a heavily bombarded, gardened, agglutinized, and homogenized regolith where the minimal iron present is in the metallic form not observable in this spectral region. However, the combination of very effective mixing and/or the occurrence of typical variations on scale much smaller than the resolution available for ground-based observations could easily have made the detection of distinct 'signatures' of typical surface features impossible.

Work with lunar soils has indicated that agglutinates tend to have experienced loss of iron and titanium relative to other soil components (Papike et al, 1982) and that the presence of reduced (metallic) iron is crucial in forming darkened layers on soil grains (Hapke et al, 1975), which is not observed to a great extent on Mercury. In fact, many workers (McCord and Clark, 1979; Vilas et al, 1984; Blewett et al, 1997) have considered the closest analogue for Mercury's surface to be lunar highlands, based on analogous slope and lack of absorption features in near IR spectra. Spectral reflectance spectra centered over plains have often indicated a shallow absorption feature associated with iron in pyroxene, possibly similar to the iron feature in Apollo 16 lunar soil (Vilas et al, 1988). On the other hand, an absorption feature associated with plagioclase at 1.25 microns has not yet been observed in near IR spectra of Mercury (Blewett et al, 2002). Of course, Mg-rich versions of pyroxene or olivine with some iron can't be ruled out, nor can metallic iron (Rava and Hapke, 1987; Clark and McFadden, 2000; Burbine et al, 2002)). Some workers have suggested that the maturity-iron relationship developed based on laboratory measurements for lunar soils (Lucey et al, 1995) doesn't hold for the regolith of Mercury (Shkuratov et al, 1999; Warell, 2001; Hapke et al, 2001) or, for that matter, for any regolith with iron-bearing constituents other than pyroxene (Clark and McFadden, 2000). Nor are the spectra consistent with high metallic iron abundance, which would create lower slopes and reflectivity (Hapke, 2001).

If Mercury has meteoritic input of metallic iron comparable to the Moon's (0.5%) if its regolith is truly like the lunar highlands, and if the maturation process, involving darkening due to submicroscopic iron production, is comparable to the Moon's, then Mercury, like the Moon, would have a silicate iron content of a couple of percent (Hapke, 2001; Warell, 2002). Are these assumptions all true? The higher vapor production and sputtering rates on Mercury predicted by Cintala (1992) would modify the maturation process, darkening, and submetallic iron production considerably. Even if impacts occur with the same frequency and scale, the impactor population could have a different composition and proportional contribution to metallic iron abundance. In other words, when it comes to predicting abundances of iron components in the regolith, direct comparison between the Moon and Mercury may be fraught with difficulties. All evidence points to relatively low iron abundance, and the meteoritic contribution alone would provide a minimum estimate. So, with no information available on systematic differences in the composition of impactors, the lunar highland abundance estimate would provide a minimum likely iron abundance with a large errorbar.

Although the presence of Mg-rich and Fe-poor mafic minerals would be consistent with an enstatite chondrite origin for Mercury described above (Wasson, 1988), high density and the presence of a partially molten core is not consistent with this model (Consolmagno and Britt, 1988). However, volatilization of the silicate mantle would produce an enstatite achondrite, or aubrite, composition containing Mg-rich pyroxene plus metallic iron and sulfur accessories which could be constrained to match the observed density. On the other hand, the spectral feature associated with sulfur at about .5 microns is not observed in Mercury spectra, although it appears in the spectra of enstatite asteroids. Also, the Mercurian albedo is half what would be anticipated and is observed for enstatite asteroids.

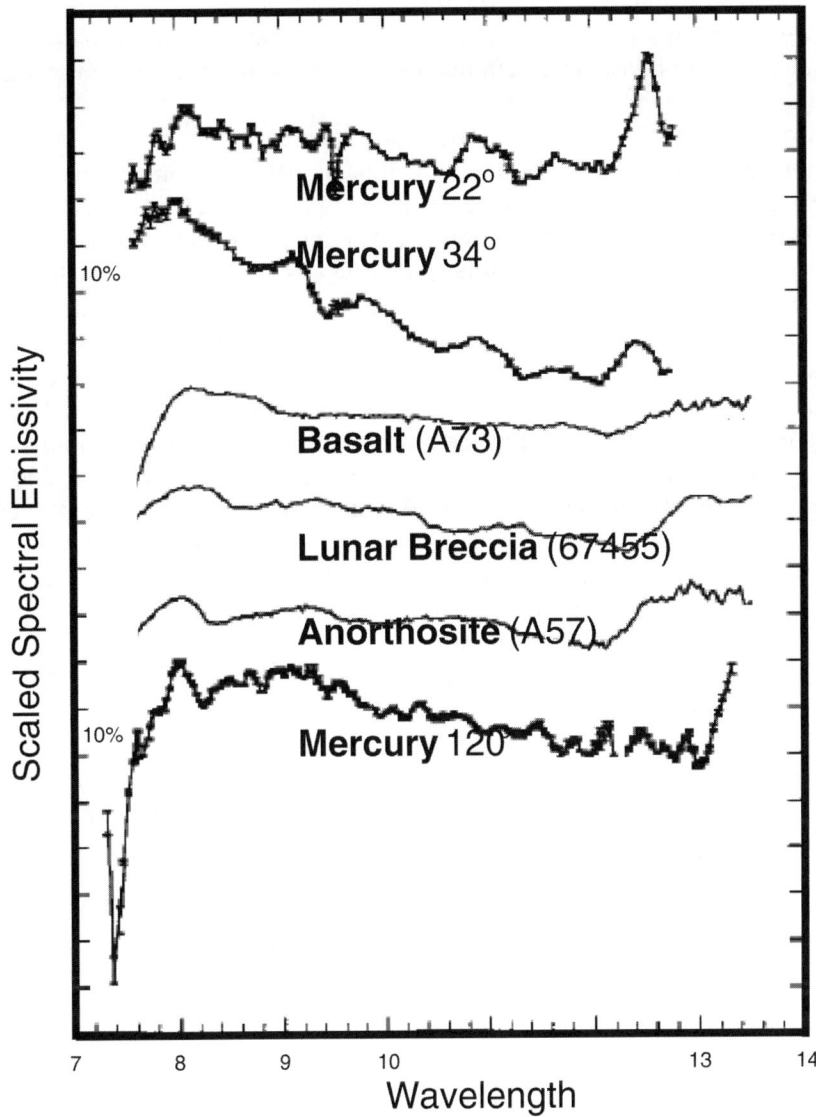

Figure 4-9. Comparison of ground-based Mid IR spectra of Mercury to known lunar minerals: Note variation in Mercury spectra, and by implication in geochemical signatures of surface terranes, in one case indicating closer kinship to lunar basalt, in other to lunar anorthosite. (Reprinted from Sprague et al, 1994, with permission from Elsevier.)

Ground-based mid infrared observations (**Figures 4-9a, 4.9b,** and **4.9c**) have, rather recently, provided new insights concerning Mercury's surface mineralogy (Sprague et al, 1994; Sprague and Roush, 1998; Sprague et al, 2002 and Cooper et al, 2003) at resolutions of hundreds of kilometers. All of these workers found that the regolith compositions derived are consistent with observations indicative of the presence of intermediate to mafic rocks which are more feldspathic than the lunar regolith, and very low in oxidized iron.

Comparisons have been made with laboratory spectra of potential components predicted by the range of models of formation (Sprague et al, 1994, Sprague and Roush, 1998; Cooper et al, 2003a, 2003b) discussed in the Interior section, including enstatite or olivine rich compositions predicted by model which require enstatite chondrite (Wasson, 1988) or chondrite (Wetherill, 1988) for initial compositions, a range of feldspars, felspathoids, and alkali basalts predicted by models which require volatilization of the crust and outer mantle (Cameron, 1988), plagioclase and pyroxene bearing lunar breccias and basalts for comparison to lunar magma ocean derived constituent mixtures, and refractory-rich spinels predicted by the equilibrium condensation model (Lewis, 1988). Cooper and coworkers (2003a, 2003b) have not been able to verify the presence of particular mineral constituents because they have not observed discrete spectral features that are clearly associated with specific minerals. However, they have observed large-scale spectral features suggestive of lunar breccias and a lack of contrast associated with a particulate regolith.

On the other hand, Sprague and coworkers (1998) have correlated the spectral features of some of these components with features in Mercury spectra. Spectral features in the 2.5 to 7 micron range, where volume scattering dominates, indicate the presence of the minerals iron poor pyroxene, such as hypersthene or diopside, and sodium rich plagioclase, such as labradorite. These spectra show the high emissivity attributed to the effect of strong thermal gradients on grains of 30 micron average diameter in a vacuum (**Figure 4-10**). In the region above 7 microns, dominated by fundamental molecular vibrations for minerals and revealing bulk rock type, the closest matches appear to be iron poor basalt and labradorite (sodium-rich plagioclase) for most locations, with some evidence of ultrabasic mineral assemblages in certain areas. Universal plagioclase features are centered near 8 microns, the Christiansen frequency, and 12.5 microns with exact positions indicative of Na substitution for Ca. A broad hump seen in plagioclase occurs from 8 to 10 microns. Mercury spectra show a range of sodium abundances. The olivine or pyroxene feature around 10.5 is seen in spectra to varying degrees. Declining symmetry in this feature is indicative of the degree of substitution of iron for magnesium. Some spectra show similarities to spectra of lunar anorthositic breccias, others to mixtures of plagioclase and pyroxene with more alkalic compositions. No indication of spinel or feldspathoid can be seen. These assemblages, although not resembling any chondrite, may have similarities to enstatite achondrite, or aubrite, which has plagioclase, low-iron pyroxene, and little metallic iron. All spectra show variations in sodium abundance. No one spectrum contains

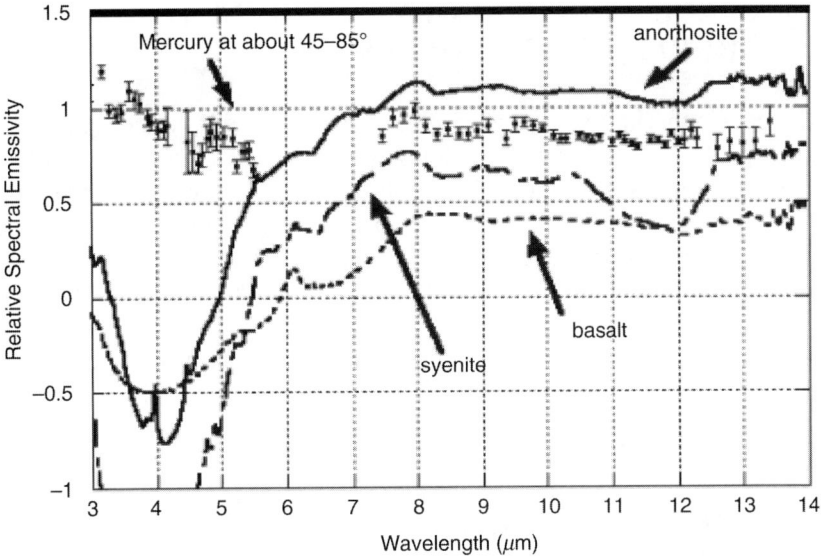

Figure 4-10. Effect of High Thermal Gradient on Mid-IR Spectra. Henderson and Jakosky (1997) illustratre that emissivity, and thus IR spectra, vary as a function of grain size. Emissivities in the mid IR region are significantly higher for larger grain sizes, which may be why Mercury spectra show higher emissivities than laboratory spectra of known mineral rock powders (From Sprague et al, 2002). Reproduced with permission of AGU.

all of the features of felsic or mafic components, suggesting mixtures of compositions. Radar transparencies for Mercury's surface are also indicate of relatively low abundances for reduced, or metallic, iron, which would not be observable with this, or near infrared spectral, techniques.

The combined spectral evidence of relatively low FeO and the presence of alkalis, as well as the spectral variability in terranes on the basis of visible and infrared properties, has provided support for the hypothesis that the plains are early and late stage iron-poor differentiates deposited as flood basalts (Jeanloz et al, 1995 and others). Intercrater plains would be the earliest Mg-rich basalts associated with crustal expansion. Smooth plains would be later alkali rich basalts associated with basin formation. Indirect confirmation for the basaltic compositional character of Mercury's smooth plains has come from comparisons of radar transparency of this terrane to lunar maria (Harmon, 1997). On the other hand, most of Mercury's regolith is typically 2 to 3 times more transparent than lunar maria to microwave radiation, indicating much lower FeO and TiO2 overall and a lack of widespread lunar style ferromagnesian volcanic activity (Taylor, 2001).

Mercury rocks are thus likely to contain less Fe and Ca but more Mg and Na than lunar rocks. For purposes of comparison, typical lunar rock suite compositions are plotted in **Figure 4-11**. No clear analogues with simultaneously high Mg and Na are found among lunar rock suites.

4.4   SPACE WEATHERING AS A REGOLITH MODIFICATION PROCESS

Space weathering, as a surface modification process, is the maturing, or darkening of regolith surface particles which occurs as a function of age. Such weathering is presumed to be induced by both micrometeorite impact and solar wind sputtering. According to the widely held vapor deposition model, both processes induce the production of a thin (angstroms) layer of submicroscopic metallic iron (SMFe) particles around regolith grains containing iron-bearing silicates (Hapke, 2001). This model was developed based on observations of lunar regolith materials. How well it works on Mercury depends on Mercury's surface composition, which is highly unconstrained. However, Hapke (2001) calculated the amount of SMFe which would be required to produce the observed variations in albedo, folding in the increased rate micrometeorite impact estimated by Cintalla (1981). Overall iron abundance in the crust could be constrained based on the SMFe required to generate the observed albedos. On this basis, a low Fe value (2 to 6% Fe) was predicted for the Mercury crust. This prediction would tend to support a selective accretion or post-accretion vaporization model for Mercury's origin and/or iron-poor basaltic volcanism.

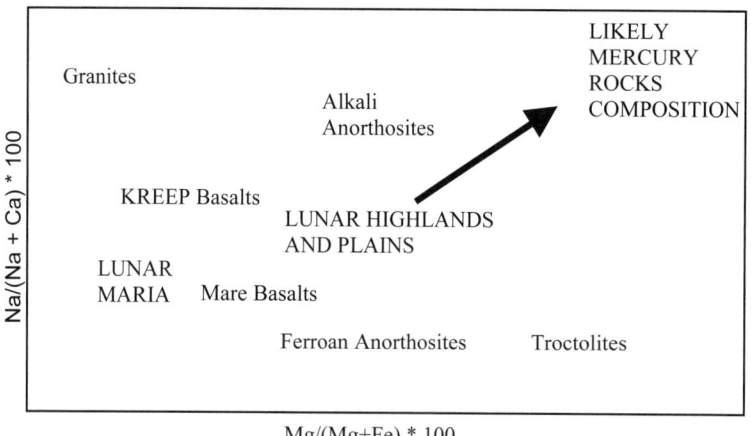

Figure 4-11. Mercury Rock Compared to Lunar Rock Composition. There are high Na lunar rocks and high Mg lunar rocks. The closest compositional analogue to Mercury rocks are alkali anorthosites, but there are no lunar rocks with the combined high Mg and Na abundances which might be found in Mercury rocks.

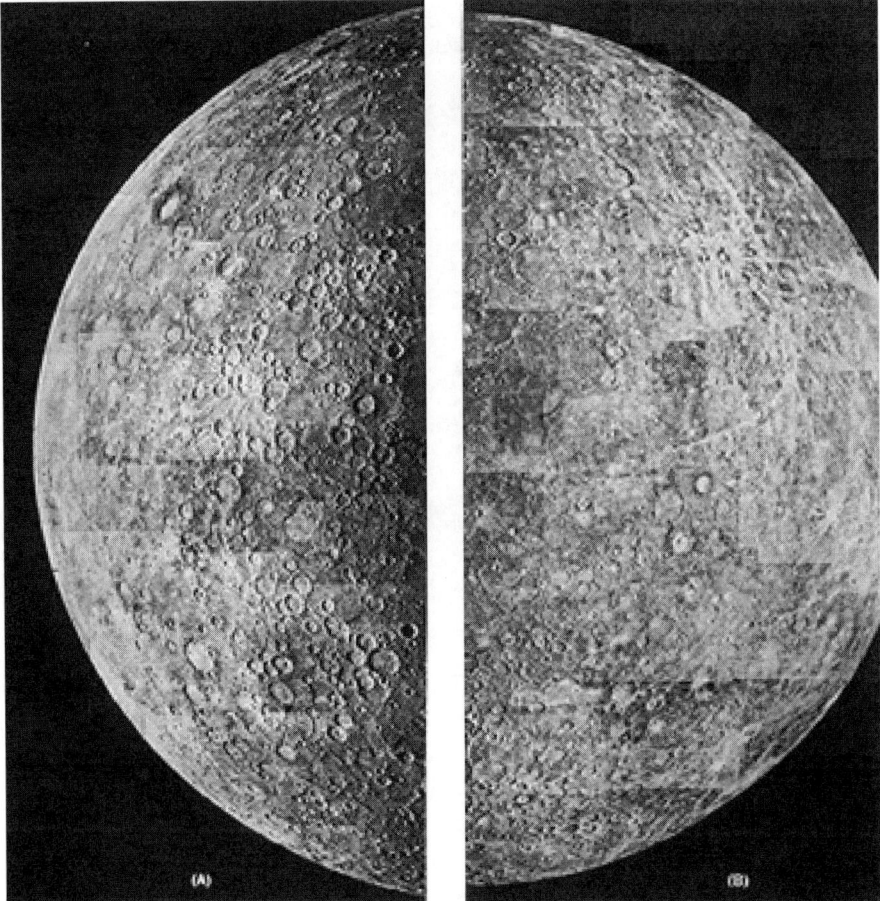

Figure 4-12. Mariner 10 mosaics Showing Major Terranes in eastern and western quadrants for imaged hemisphere, showing the crater as the ubiquitous landform from the NASA Atlas of Mercury. The most ubiquitous terrane here is plains, varying in albedo and extent of cratering, and not highlands or heavily cratered terrane as on the Moon.

4.5   THE NATURE AND COMPOSITION OF MAJOR TERRANES

Three major terranes, or geochemical provinces, were observed on Mercury's heavily cratered surface by Mariner 10 (See **Figure 4-12**). Distinct boundaries associated with intercrater plains, small volcanic deposits, and recent impact ejecta have been identified on the basis of albedo and color ratio in the recalibrated Mariner 10 images (Robinson et al, 1997). Early on, Mercury's heavily cratered surface was inevitably compared to the lunar surface, which had been recently visited.   On the other hand,

Figure 4-13. Typical Heavily Cratered Terrain from the NASA atlas of Mercury. Note this oldest terrane on Mercury is still not as uniformly bright as the lunar highlands.

Mercury's surface is much smoother on a kilometer scale (Clark et al, 1987), with 2 to 3 kilometers elevation difference between lowlands and highlands as compared to 7 kilometers on the Moon. Of course, Mercury has many features for which there are no lunar analogues. In addition, Mercury is shockingly more Earthlike in many ways. A comparison of the features of Mercury in the context of the Moon and the Earth can be seen in **Table 4-2**.

Small patches of Heavily Cratered terrane (**Figure 4-13**), analogous to, although not as extensive as, the lunar highlands, were observed. The lunar highlands are thought to be remnants of early crust formed from a magma ocean.

Smooth Plains (**Figure 4-14**) have most frequently been associated with Caloris basin and its antipode, although similar terrane is observed in smaller basins of comparable age, including Beethoven. This terrane was thought at the time to be analogous to, although not as extensive as, the lunar maria, which of course has a volcanic origin, or possibly to be analogous to the less extensive lunar plains deposits, of more controversial origin. Smooth plains have a smoothly undulating, down-bowed character and ridges analogous to lunar maria (Harmon, 1997), and a mare-like roughness.

The most extensive terrane, for which there is no clear lunar analogue, is the Intercrater Plains terrane (**Figure 4-15**). These plains are in some cases associated with heavily degraded basins, including Dostoevskij and Tolstoj,

Figure 4-14. Typical deposits of smooth plains in the interiors of basins from the NASA atlas of Mercury. Note darker, smoother, less cratered, and more uniform appearance. This terrane is the closest analogue to volcanic plains.

and cover decreasing area with decreasing age. (Uchupi and Emery, 2003). Currently, despite earlier controversies, a volcanic origin is preferred for the Intercrater Plains, not only because of the difficulty of generating the extensive volume of material involved as impact ejecta, but also because of the association with a period of potential opening of conduits for flood basalt associated with core formation and crustal expansion. Now, a range of evidence from ground-based spectrometry and detailed study of morphology, described below, appears to support a volcanic hypothesis. Seismically-induced swirl features have been observed antipodally to Caloris, as they have been observed antipodally for some basins on the Moon (Uchupi and Emery, 2003).

Visual comparison of prominent plains on the Moon, Mars, and Mercury illustrates the differences between volcanic plains on the three planets

(**Figure 4-16**). Note the differences in roughness and apparent modification after formation, on Mars by wind, on Mercury by scarp formation.

Table 4-2. **Mercury in Context of Other Terrestrial Planets**

Planet	Moon	Mercury	Earth
Surface	Regolith	Regolith	Complex
Atmosphere	Thin exosphere	Thin exosphere	Thick atmosphere
Impact Events	Heavily Cratered Impact-controlled surface processes including basin formation	Heavily Cratered; Impact major role in surface processes including basin formation; also major tectonic episodes	Little evidence of cratering, surface processes not impact controlled
Tectonism	Magma Ocean Formed crust still in highlands, inactive for almost 3 billion years	Tidal despin, Core formation and solidification led to tensional faulting lineaments, massive resurface of original crust, and network of scarps	Plate tectonics much activity at plate boundaries
Volcanism	Mafic flood basalt filled impact basins less extensive highland volcanism	Intercrater plains extensive volcanic flood basalts during core formation, later some lunar mare style volcanism	Extensive volcanism at plate boundaries
Crust	Asymmetric Thick magma ocean formed crust covers much of surface	Thick, asymmetric, patches of magma ocean formed crust, mostly heavily resurfaced	Thin, mantle recycled plates, heavily resurfaced
Core/Magnetism	Small, solid core Remnant fields	Large core probably partially fluid and magnetic dipole	fluid core with dynamo and magnetic dipole
Polar Deposits	volatiles mixed with regolith	Clearly solid volatile layers in cold spots	Extensive ice deposits

Table 4-3. **Radar Features in the unimaged hemisphere**

Location (Long/Lat)	Feature
Poles	Thick volatile deposits in permanently in shadowed craters
Antipode to Caloris	Chaotic terrain related to Caloris basin formation
345W/55N	Bright ring dark floor Volcanic Caldera Complex
345W/29S	Recent Impact Crater Complex
300W equatorial	Shield volcano complex
220-250W equatorial	Rough Volcanic Plains
240W/17N	Bright ring dark floor volcanic Caldera

What does the unimaged hemisphere hold? Radar observations may indicate the presence of greater topographic variation, regions of great basins or extensive cratering, and roughness, with plains of increasing roughness in the equatorial region from 220 to 250 degrees longitude possibly associated with a volcanic feature (Zohar and Goldstein, 1974; Clark et al, 1988; Butler et al, 1993) (**Table 4-3**).

Figure 4-15. Typical intercrater plains from the NASA atlas of Mercury. Note the variable character of this terrane in terms of albedo, roughness, crater density. Note suggestions underlying structure.

4.6 CONCISE SUMMARY OF MERCURY'S GEOLOGICAL HISTORY

A milestone chart of Mercury's geological history (**Figure 4-17**) illustrates the major events which are thought to have occurred relatively early in Mercury's history and the uncertain time frames of these events.

The available evidence indicates that, early in Mercury's history, tidal spin-down caused lineament formation (Melosh and McKinnon, 1988). Early in its history, Mercury, along with the other terrestrial planets, then experienced a period of intensive impact activity, known as the 'late, heavy bombardment'. Near the end of the bombardment, a period of crustal expansion at the onset of core formation, expressed as orthogonal surface strike slip faulting and ridge and trough formation occurred, along with large-scale interior melting and magma flow through fractures to form the inter-crater plains (Spudis and Guest, 1988; Clark et al, 1988). Intercrater

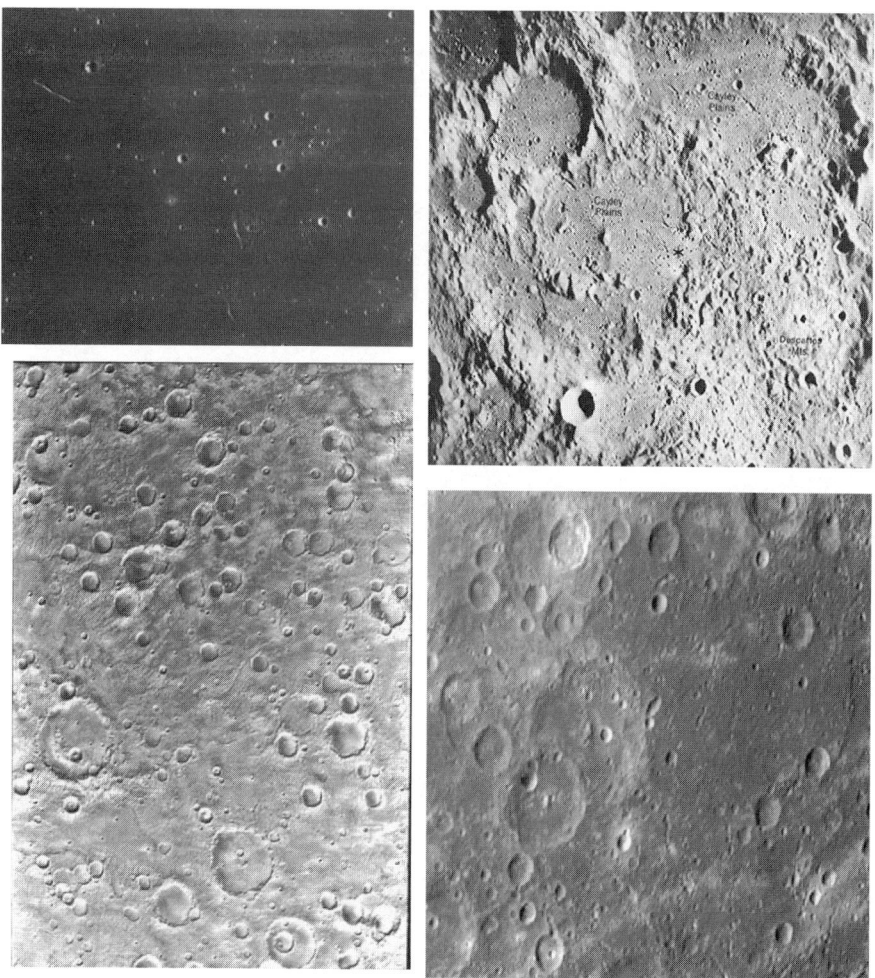

Figure 4-16. Comparison of Moon, Mars and Mercury plains (NASA photographs). Top right represents lunar maria, clearly volcanic plains. Top left represent lunar 'plains', brighter, less continuous, eroded by impact. Bottom right represents Martian plains, extensively smoothed by wind erosion. Bottom right represents Mercury intercrater plains, showing evidence of impact and perhaps volcanic flooding, of more controversial origin.

plain formation eventually stopped as partial shrinking and solidification of the core resulted in crustal compression, the formation of surface thrust scarps, and the closure of surface fractures (Spudis and Guest, 1988). Magmas prevented from erupting directly during this process may have undergone extensive differentiation to appear, for example, as alkali-rich feldspathoid basalt (Jeanloz et al., 1995). Alternatively, intercrater plains may constitute impact ejecta from late heavy bombardment analogous to the lunar Cayley formation (Wilhelms, 1987). Smooth plains probably formed as lunar-like flood basalts in the impact basins (Spudis and Guest, 1988).

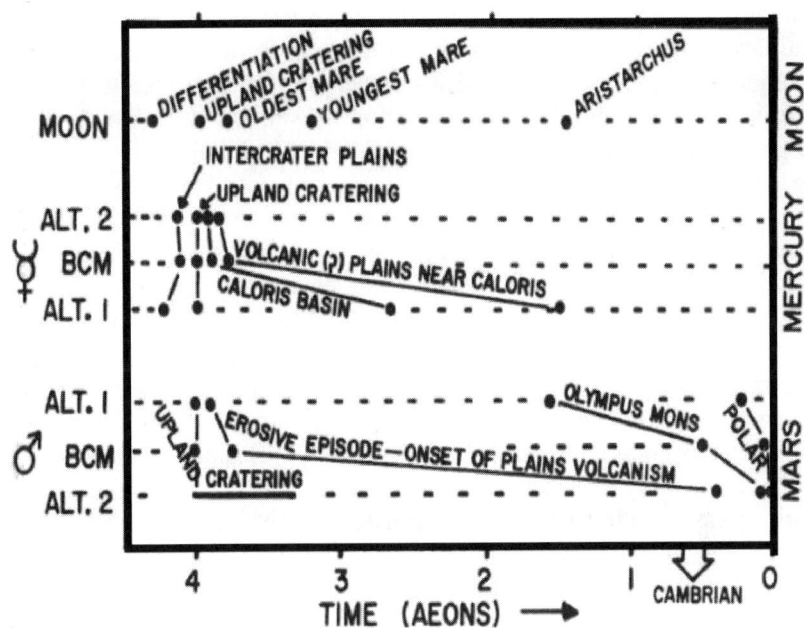

Figure 4-17. Comparative chronologies for Mercury, the Moon and Mars (Chapman, 1976), based on alternative 1 (Murray et al, 1975) and the working model, BCM, and alternative 2 proposed by Chapman. Note very different ages for most recent volcanic plains formation and by implication the rate of core cooling. (Reprinted from Chapman, 1976 with permission from Elsevier.)

4.7   IMPACT ACTIVITY AND CHRONOLOGY

Mercury experiences more energetic bombardment by asteroids and comets traveling close to the sun than the Earth. Although there are fewer Mercury-crossing than Earth-crossing asteroids, asteroids and comets within Mercury's orbit have greater confluence.

Knowledge of the **cratering** record on large silicate bodies, without the complicating effects of water and/or wind weathering, is restricted to the Moon and Mercury. Gravity-inferred morphology and spectrally-inferred mineralogy extrapolated from studies of the Moon, have improved our understanding of excavated substrates and of the cratering process as well. Apparently anorthositic signature of impact ejecta on the recalibrated Mariner 10 images may indicate the presence of an underlying anorthositic crust on Mercury (Robinson et al, 1997) as well as the Moon.

Craters are the most ubiquitous landforms on Mercury, covering the entire observed surface. Population statistics for the smaller, and younger, craters are still unknown, due to the lack of coverage, particularly at high resolution, by Mariner 10 (the best resolution was 100 meters, typical resolution hundreds of meters, for the one hemisphere observed). Comparable numbers of the largest, most complex impact features, multiring basins, exist on the Moon and Mercury; however, a dearth of crater less than tens of kilometers in size is thought to be due to obliteration by subsequent plains formation. The greater degradation and virtually non-existent debris fields of these large, Hermean impact structures are consistent with gravity scaling. The smaller diameter at which the transition from simple bowl to complex structures with interior terraces and peaks occurs and smaller depth to diameter ratios are trends consistent with gravity scaling as well (Smith, 1976). On the other hand, craters on both the Moon and Mercury show well developed bright ray systems (Slade et al, 1992; Harmon and Slade, 1992).

Comparative studies of the morphology of impact structures on the smallest terrestrial planets has provided some insight into the nature of Mercury's crustal processes (Potts et al, 2002). Basins on the Moon and Mercury apparently experienced an analogous sequence of events, scaled by gravity, following formation of a transient cavity, which induced uplift, followed by isostatic relaxation, and volcanic infilling. Multi-ring basins on all three bodies have a similar ring spacing (Potts et al, 2002). Based on the size of transient cavities and density of impact basins (Melosh, 1982), Mercury's early crust was most like the lunar farside, but more viscous than the lunar or martian crusts (Potts et al, 2002). Because it was also hotter, a dryer crust is also implied.

The cratering record has been used extensively as a chronometer, a means of providing relative ages for surface strata. Such ages are based on the accepted model for the size and numbers of projectiles in the neighborhood of the terrestrial planets as a function of time. The similarity in shape of plots of crater diameter versus density on the terrestrial planets apparently indicates similar impact history (**Figure 4-18**) (Strom and Sprague, 2003). The average size and abundance of potential impactors have decreased as a function of time.

Based on crater density versus diameter relationships, Mercury's surface is old. Observed surface structures can very generally be classified as those in high crater density regions which were resurfaced by intercrater plains, and those in lower surface density regions postdating intercrater plains formation. Evidence for an early episode of intensive impact activity which occurred soon after planet formation may be observed on all the planets in the inner solar system, including Mercury (e.g., Neukum et al, 2001). This period of post-accretion cataclysm is often referred to as the late heavy bombardment (Wetherill, 1988). The question remains as to how cataclysmic this activity was, and, if it was not cataclysmic, whether the same population of large impactors was involved simultaneously for all of the inner planets, or whether discrete impactor populations were involved either additionally or exclusively for each planet (Chapman, 1976).

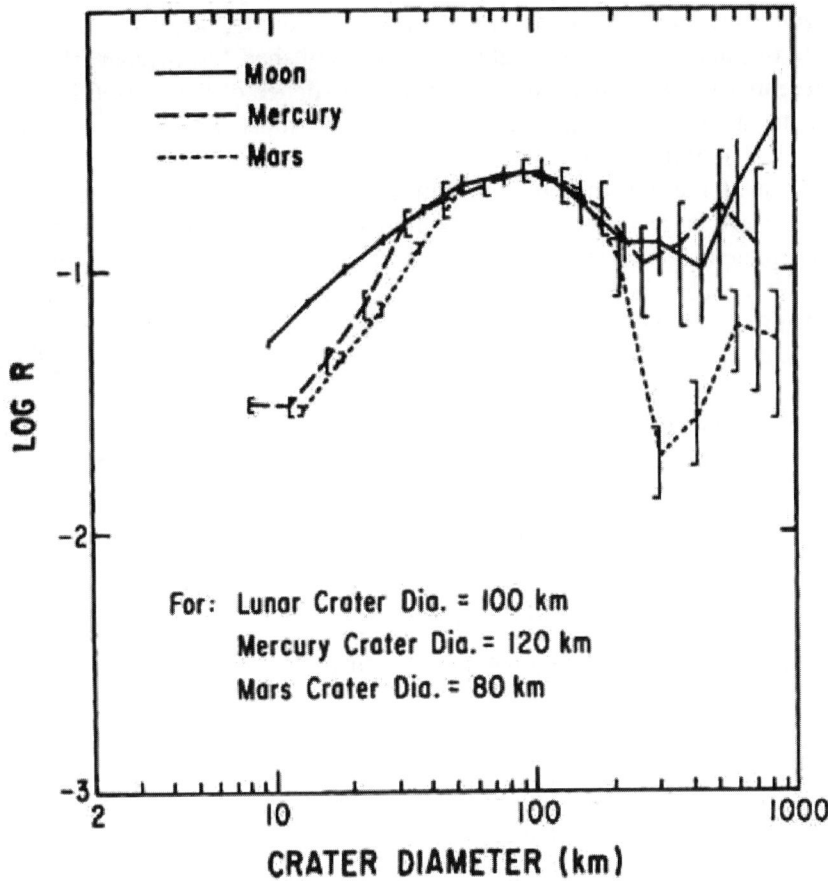

Figure 4-18. Crater Size/Frequency Distributions for the highlands of the Moon, Mars, and Mercury for craters greater than 150 km diameter. (From Strom and Neukum, 1988, The cratering record on Mercury and the origin of impacting objects, in Mercury, Copyright 1988, The Arizona Board of Regents. Reprinted by permission of the University of Arizona Press.)

In the case of Mercury, a population of Mercury-crossers known as vulcanoids has been postulated, but has not been found due to the difficulty in observing such a population (Leake et al, 1987). Despite the lack of samples from Mercury, if the same population bombarded all of the planets, then an absolute chronological timescale can be established for dating terranes on the Moon and Mercury, where comparably heavily bombarded terrane have not been completely buried. If separate planet-crosser populations were involved, only a relative stratigraphically based timescale can be established for Mercury.

The consensus is that the impact cratering rate is comparable on the Moon and Mercury (Neukum et al, 2001). Even though the population of late heavy bombardment impactors is predicted to be somewhat smaller in scale at Mercury's orbit, higher gravity produces features larger in scale for a given impactor. On this basis the timing of the late, heavy bombardment period of impact activity is relatively well established and comparable on all of the inner planets (Neukum, 2001); thus, evidence of that period has been used to approximate the timeframe of the most extensive period of geological activity on Mercury.

Much of the evidence of the bombardment, including the larger, older craters, were largely, although not completely, superposed by the Intercrater plains, indicating that intercrater plains formation, and the core formation process which caused this activity, occurred near the end of the bombardment. The Intercrater Plains (**Figure 4-15**) themselves are densely impacted with smaller craters, far moreso than the Smooth Plains (**Figure 4-14**), which, as the name implies, are far less covered with craters, and the average crater size is smaller. The Smooth Plains are therefore younger. The Heavily Cratered terrane (**Figure 4-13**) is saturated with larger craters, as the lunar highlands, is therefore the oldest terrane, and likely to be remnant of the earliest crust not resurfaced with Intercrater plains.

Despite these generalities, the factors of two or more uncertainty in the timescale of formation for major terranes and geomorphological features on Mercury remain (Clark, 1976; Neukum et al, 2001), due to (a) the divergence of timescales predicted by unconstrained thermal models; (b) the lack of availability of relative age determinations using characteristic crater degradation features (e.g. depth/diameter ratios) because of lack of coverage and relatively poor quality of imaging (resolution poorer than 100 meters), and 3) the uncertainty of crater production rates.

Nevertheless, Mercury features overlapping crater populations which can be used to determine a relative timescale for major terrane formation (Neukum et al, 2001). **Table 4-4** is such a time stratigraphic table (Spudis and Guest, 1986, 1988). The large, severely degraded cratered areas known as the heavily cratered terrain formed when large impactors formed many large basins and saturated already heavily cratered surfaces. This terrane still has a lower crater population density and somewhat lower albedo than the lunar highlands. Early in the Pre-Tolstojan, core formation initiates crustal expansion-induced eruption and emplacement of intercrater plains.

Throughout the pre-Tolstojan and even into the Tolstojan period, Heavily Cratered Terrain and multi-ring basins are embayed (surrounded) by intercrater plains, a ubiquitous stratum with comparable albedo but different morphology from Heavily Cratered Terrain. These plains form extensive rolling hills with a large range of crater densities, but generally fewer and smaller craters, indicating gradual emplacement subsequent to late heavy bombardment. Current observable larger basins, Dostoevskij, Shakespeare, Homer, and Surikov, were emplaced after the bulk of intercrater plains formation, during the later Tolstojan period. Intercrater plains transition into the youngest Smooth Plains during the Calorian period. Smooth Plains have a somewhat lower albedo and lower crater population density, and occur patchily in the highlands, in and around large basins and near the poles. The Calorian system formed during this period, including Caloris basin itself, and the associated Caloris Montes, Nervo, Odin, and Van Eyck formations. Scarp formation occurs during the Calorian period. Calorian two-ring basins include Rodin, Michelangelo, Bach, and Strindberg. Smaller craters formed during the Mansurian and Kuiperian periods.

Table 4-4. **Stratigraphic History of Major Geological Units (Neukum et al, 2001; Wagner et al, 2001)**

System	Major Units	Age Ga	Lunar Eq
Kuiperian	Crater materials	<3.0	Copernican
Mansurian	Crater materials	3.0-3.9	Eratosthenian
Calorian	Caloris Group (including secondaries, ridges	3.9	Imbrian
	Caloris textured plains		
	Caloris (smooth) Plains		
	Crater and Small Basin Material		
Tolstojan	Goya Formation (Tolstoj Basin deposits)	3.9-4.0	Nectorian
	Plains		
	Small Basin Plains		
	Crater materials		
Pre-Tolstojan	Intercrater plains	4.0-4.2	Pre-Nectorian
	Multi-ring basins		
	Crater materials		
Highland	Heavily cratered plains	>4.2	Highland

The general morphologies of craters formed in Mercury plains and the lunar maria are similar, and all are different from morphologies of lunar highland craters, implying that target material and not just gravity is an important factor in the impact process. On the Moon, this difference could be accounted for by the difference in structure between the highlands, composed of highly fragmented rock, and maria, composed of more coherent, solid rock. This would imply that Mercury plains, both smooth and intercrater, are more similar to the volcanic, basaltic maria, than the highlands. The more concentrated, higher velocity, and smaller dispersion of impact ejecta on Mercury than on the Moon can be accounted for by the higher gravity of Mercury. An exception would appear to be some of the bright rays on Mercury, which are hundreds of kilometers long. Perhaps these were generated by parabolic comet impacts (Schultz, 1988) or by reactivation of preexisting lineaments.

Caloris is the most prominent basin imaged by Mariner 10 (**Figure 4-19**). The complex network of ridges in its interior is reminiscent of lunar maria and probably also resulted from compressional stresses during subsidence of the basin floor, but are much more prevalent and with a stronger radial component. Tensional fractures, obviously occurring later during the

Figure 4-19. The Caloris Basin complex from NASA atlas of Mercury. Note radial and concentric features associated with impact basins on the Moon, as well as deposits of younger, darker plains known as Smooth Plains.

Figure 4-20. Hilly and lineated terrain occuring at the antipode of the Caloris Basin complex and clearly associated with its formation. Note the extraordinarily complex and chaotic nature of this terrain. (From NASA atlas of Mercury)

rebound and uplift of the basin floor which followed, cut across the ridges. Antipodal to Caloris is the hilly and lineated terrain, which appears similar to the floor of Caloris (**Figure 4-20**). Perhaps the terrain formed as a result of seismic waves refracted off the core during the formation of Caloris (Strom et al, 1975; Spudis and Guest, 1988). Alternatively, Caloris and the hilly and lineated terrain are subject to additional tidal stresses as the 'hot poles' which alternate being at perihelion.

Prominent features in the unimaged hemisphere have been detected using ground-based radar observations (Goldstein, 1970,1971; Butler et al, 1993).

A feature with highly radar reflective radar rim and floor observed near 345 degrees longitude and just south of the equator, is thought to be a fresh, Tycho-class impact crater (Harmon, 1997; Strom and Sprague, 2003). Enhancements in reflectivity are also seen in association with the Caloris Basin and young, fresh craters, such as Kuiper, clearly indicating small-scale structure due to the presence of impact debris on the surface.

Mercury alone among the inner planets can provide fundamental data on the effects of high velocity cometary impacts. Systematic differences in their shape and central peak structure should allow craters which are cometary in origin to be identified (Schultz, 1988). An effect of the relatively high impact velocities at Mercury could be a 4-10-fold increase in impact melt/vapor products which should be most evident in young craters, and provide a source of the volatiles found in cold traps near Mercury's poles (Schultz, 1988). Higher velocity impacts would also increase the rate of space weathering by a factor of 20 relative to the Moon (Cintala, 1981) and cause a higher overall rate of soil maturation around young craters.

4.8   VOLCANISM

The role of **volcanism** on Mercury is poorly understood and controversial. The intercrater plains, by far the dominant terrane on Mercury, were first thought to be volcanic, analogously to the much less dominant lunar 'light' plains observed at comparable resolution (Murray et al, 1974, 1975; Strom et al, 1975; Trask and Guest, 1975).

As samples were gathered from the lunar plains, many workers shifted to an impact origin for lunar and Mercurian plains as well (Wilhelms, 1976). Lunar rocks gathered from plains-associated sites (e.g., Apollo 16) were breccias little different in overall composition from highland rocks. By consensus, highlands were heavily cratered crust with no volcanic activity; thus, plains rocks must be of impact origin on the Moon. Lunar photogeological evidence alone still indicated later emplacement and embayment of earlier structures, supporting a volcanic origin. A somewhat more complex picture has emerged since then, raising the possibility once again of a volcanic origin for the lunar plains rocks, with a distinctive non-mare basalt composition (Clark and McFadden, 2000).

In fact, the lack of direct spectroscopic evidence for iron (Vilas, 1985,1988) and lunar-like basaltic volcanism on Mercury, has been thought to imply impact rather than volcanic origin for major terranes. The lack of intermediate-age craters, younger than the basins with which they are associated and yet older than overlying plains, features commonly observed in lunar maria, would eliminates the discontinuity in time between basin formation and volcanic flooding required for the volcanic hypothesis (Wilhelms, 1976). On the other hand, Mercury's high mean density and global magnetic field suggest a partially molten iron-rich core which should be accompanied by extensive volcano-tectonic activity. The extensive plains formation, partially overlain by the global compressional scarp system, and the small likelihood that impact basins could have generated the volume of impact melt required to form them, imply early differentiation and volcanism of a very different nature from lunar volcanism.

The present day consensus is that Mercurian plains have a volcanic origin (Strom et al, 1975; Murray et al, 1975; Dzurisin, 1978) due to their massive embayment of the heavily cratered terrain: resurfacing by intercrater plains covered most of the surface and destroyed existing craters smaller than 300 to 500 km in diameter. The lack of observed volcanic landforms, an argument advanced against volcanic origin, is not surprising, as, for flood style volcanism, such features would be local, small, and thus at or below the resolution of available imaging. On the other hand, the highly variable extent to which plains material are associated with similar sized basins as well as the distinctive difference in color and albedo between underlying basin and plain deposit material are both indicative of an internal origin (Trask and Strom, 1976).

The volcanic materials are unlikely to be entirely typified by lunar mafic flood basalt from deep magma sources (Jeanloz et al., 1995; Sprague et al., 1994, 1997), and not only because these plains do not have a lunar basalt like signature. Such magma production entirely from deep sources would result in extensive heat loss from the interior, and ultimately complete cooling of the core, which must still be at least partially molten. In the case of Mercury, cooling has already occurred for as much as 50% of the interior (estimated based on crustal shrinking), an amount intermediate between the geologically active Earth (10%) and the inactive Moon (65%) (Jeanloz et al, 1995). These observations do not preclude volcanism, but constrain much of the volcanic activity to the crust and upper mantle. Volcanism occurring after the crustal shrinking that generated the scarps would presumably have shut down deep heat pipe sources. The observed alkali-rich, silica-poor compositions could represent alkali lavas from secondary eruption of highly differentiated basaltic magma (Jeanloz et al, 1995). Two models to be considered for volcanism on Mercury involve either extensive deep early Mg rich flood basalt formation of the intercrater plains followed by minimal shallow late high alkali flood basalt formation of the smooth plains, or similar extensive deep early Mg rich flood basalt formation of the intercrater plains followed by partial exposure through impact of alkali rich layers down to anorthosite (Strom, 1997).

Current estimates of Hermean global volcanic productivity, and conclusions regarding eruption style and age, are likely to be inaccurate because 55% of the surface has not yet been imaged. At Earth, Venus and Mars, more than half of their volcanic sources occur in <30% of the surface area (Crumpler and Revenaugh, 1997); mare deposits cover only 17% of the Moon's surface (Head, 1975) and are not randomly located (Head and Wilson, 1992; Guest and Murray, 1976; Whitford-Stark and Head, 1977). The lack of analogous information for Mercury, has influenced global volatile estimates (an issue in modeling early Solar System formation).

The possibility that the unimaged hemisphere is a center of volcanic activity is supported by ground-based radar observations of this hemisphere. Centered at 300 degrees longitude near the equator, is a large, heavily cratered, plateau surrounded by relatively young, uncratered plains rough on a cm-size scale (Harmon and Slade, 1992). Near 345 degrees north of the equator and near 240 degrees at the equator are roughly circular haloes with dark centers (Harmon, 1997; Zohar and Goldstein, 1974)). These features have signature and size comparable to radar images of shield volcanoes on other planets. The work of other (Robinson and Lucey, 1997, 1998; Sprague et al, 1994) has supported this hypothesis, establishing that spectral heterogeneity and patterns of distribution in the intercrater plains are consistent with differences in composition of lava flows or pyroclastic centers.

4.9   TECTONIC ACTIVITY

Extensive **tectonic activity** on Mercury has produced unique, large structural features. The formation of these structures has resulted, respectively, from small changes in the shape of the lithosphere induced by the sheer from tidal despinning, expansion from core formation, and contraction from core solidification (Melosh and McKinnon, 1988). The relatively extended and subdued nature of this activity resulted from Mercury's dearth of radioactive elements inducing lower internal heating and surface heat flow combined with the planet's high dayside surface temperature inducing relaxation of topographic features (Melosh and McKinnon, 1988).

The orthogonal lineaments trending north to south, northeast and northwest, or east to west, from equatorial to polar latitudes respectively,

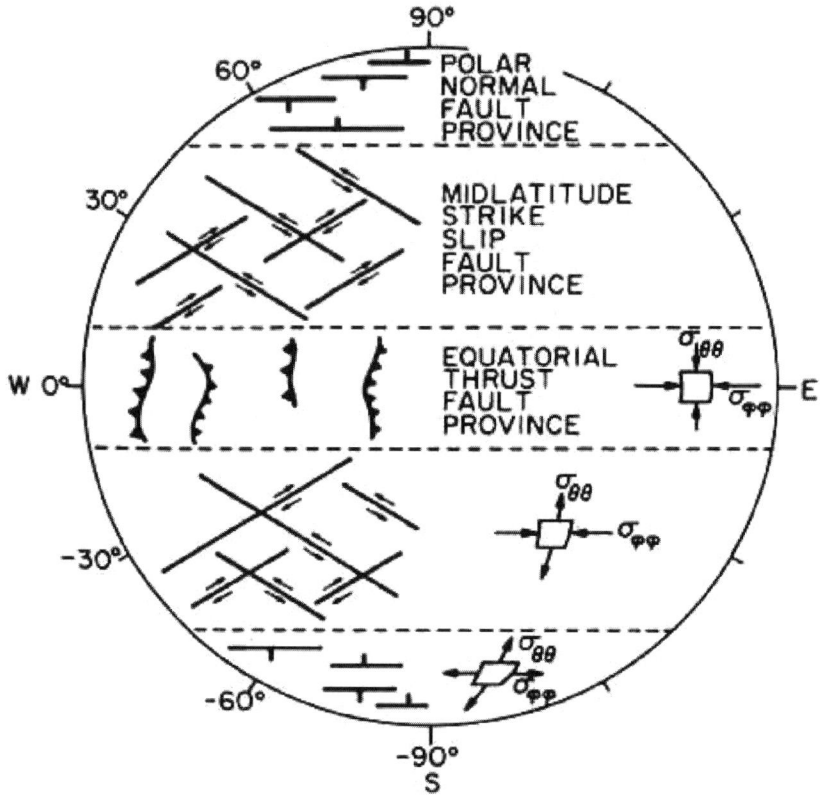

Figure 4-21. Mercury Fault Systems, including equatorial band of thrust faults (scarps) and orthogonal strike slip faults as a result of tectonic response to tidal despinning (From Melosh and McKinnon, 1988, The tectonics of Mercury, in Mercury, Copyright 1988, The Arizona Board of Regents. Reprinted by permission of the University of Arizona Press.)

which resulted from early despinning and tidal lock (Melosh and McKinnon 1988) are partially visible but have been largely obliterated. Tidal slowing would have resulted in a more spherical shape for Mercury (**Figure 4-21**), contracting the once elongated equator, and expanding the once flattened polar regions (Strom and Sprague, 2003). Mariner 10 stereo images have provided some evidence for portions of this orthogonal network of tensional features, subdued but partially reactivated by local impact events or subsequent tectonic activity (**Figures 4-22, 4.23, and 4.24**)(Clark et al., 1988; Leake et al, 1988). Further support for this model can be observed in the oldest terrane, where such trends for linear features are most frequently observed (Dzurisin, 1978; Leake et al, 1987).

The onset of core formation would have caused crustal expansion and resulted in the eruption of the intercrater plains through fractures created by the reactivation and opening of the existing early orthogonal fracture system (Clark et al, 1988; Leake et al, 1987, 1988). Evidence for such a network of

COLOR PLATES

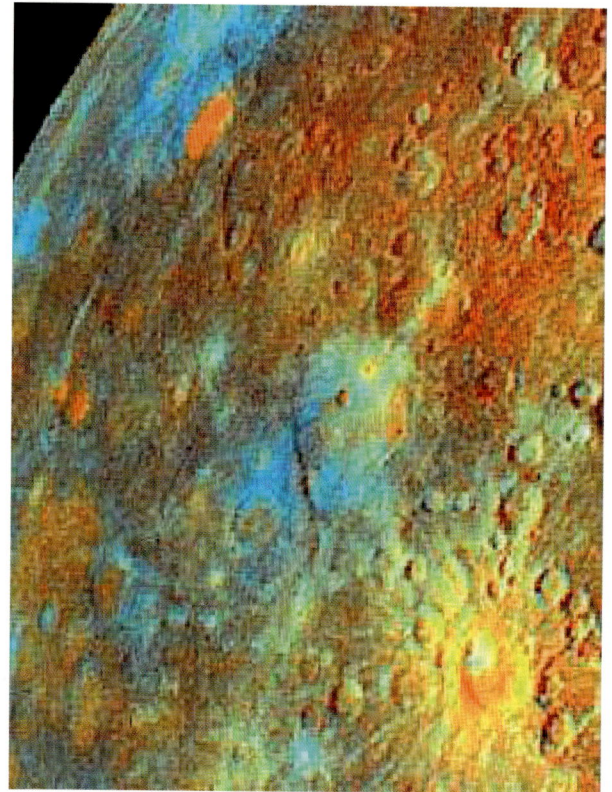

Figure 4-7.

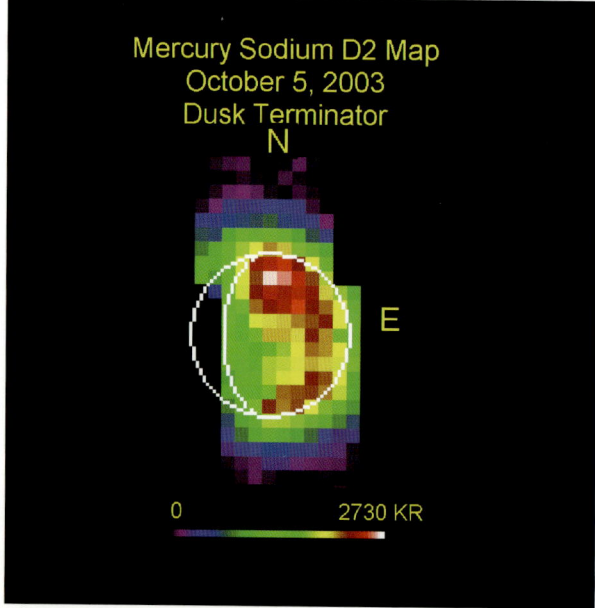

Figure 5-2.

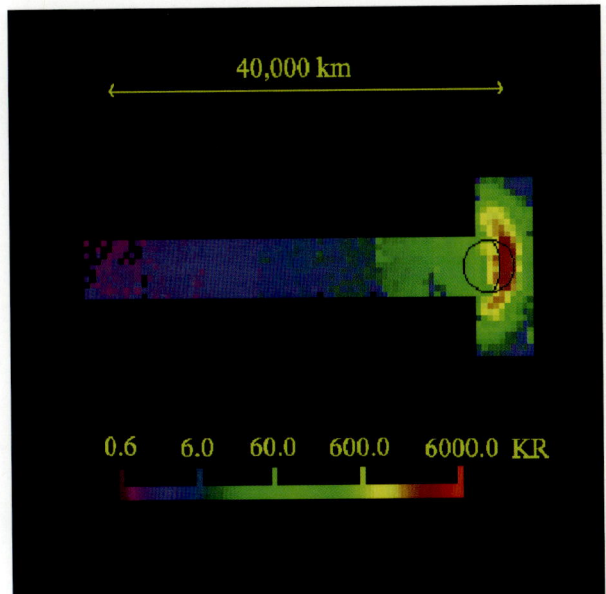

Figure 5-3.

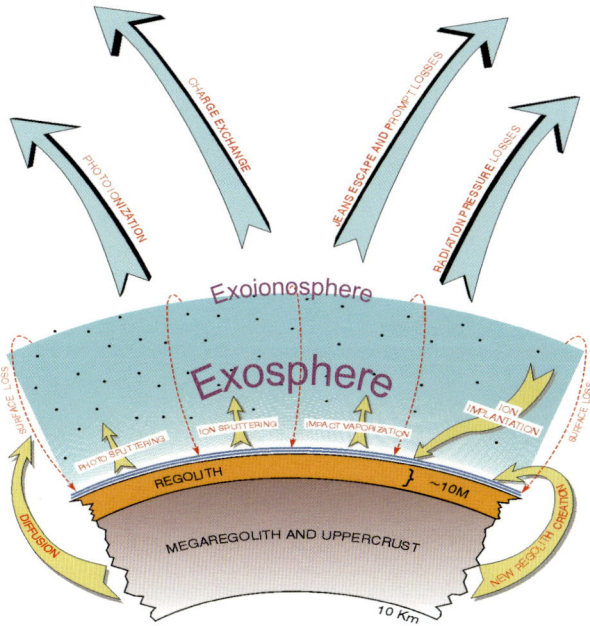

Figure 5-5.

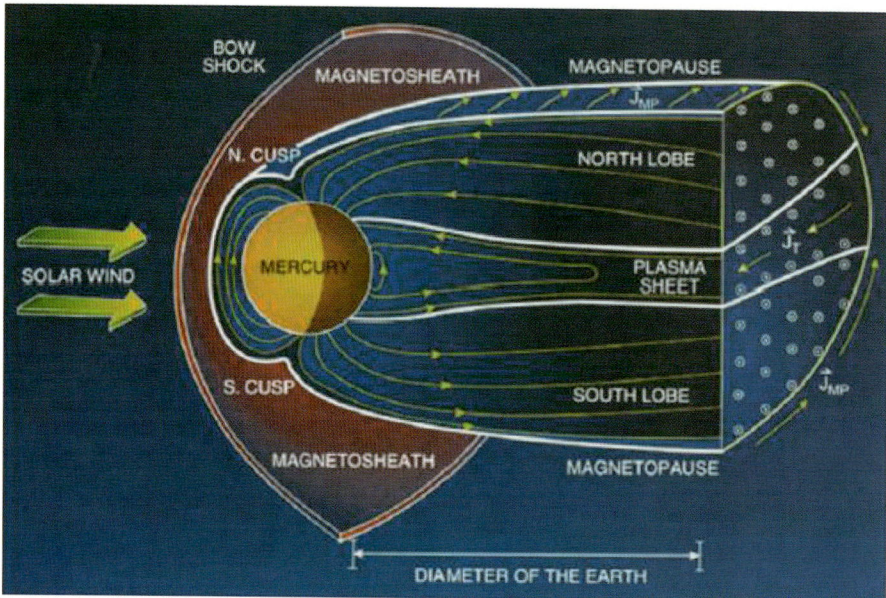

Figure 6-12.

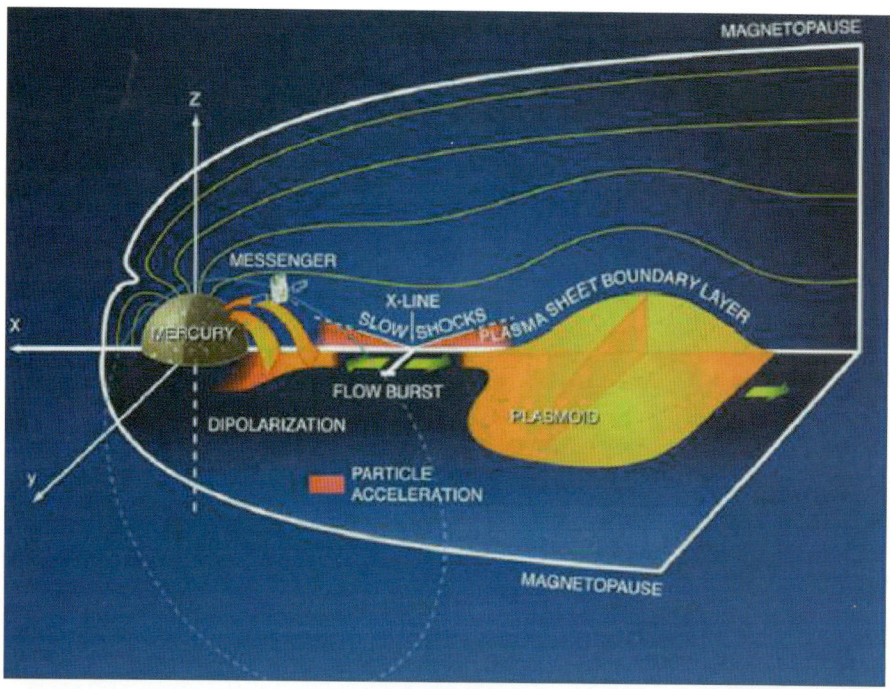

Figure 6-13.

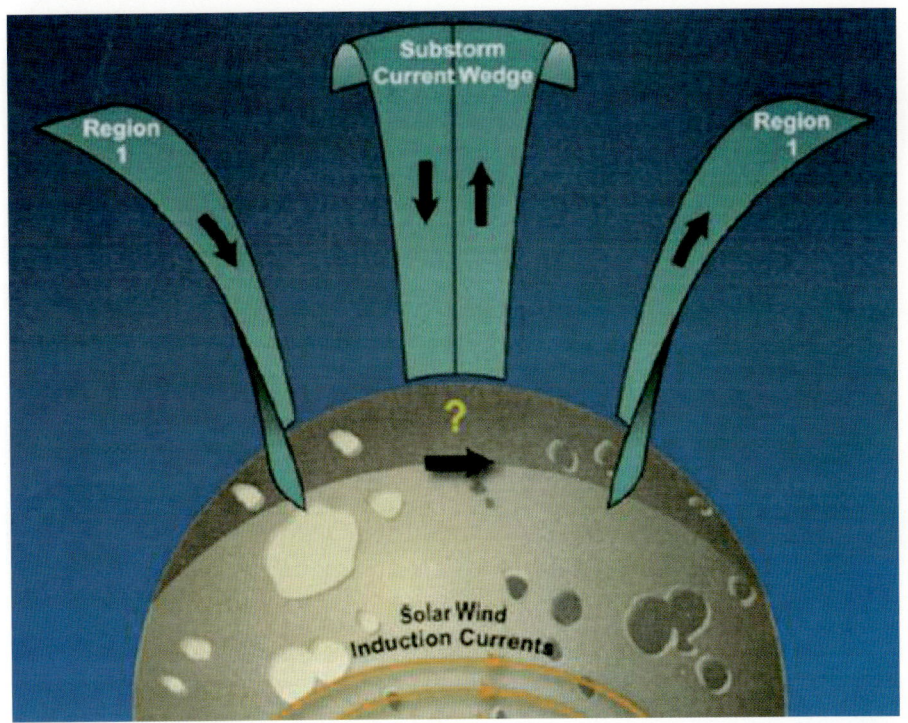

Figure 6-17.

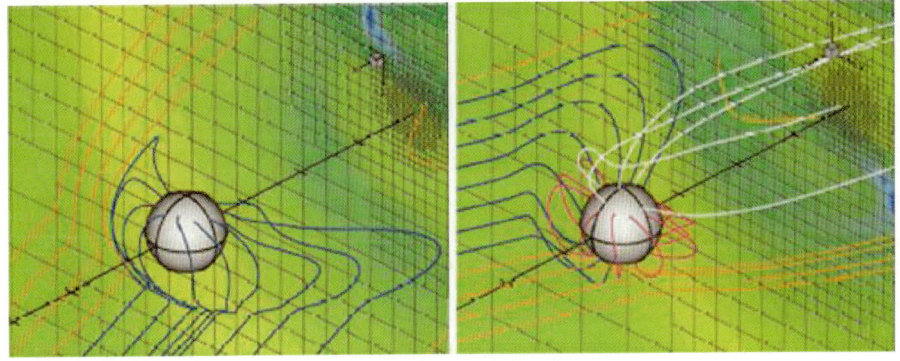

Figure 6-18.

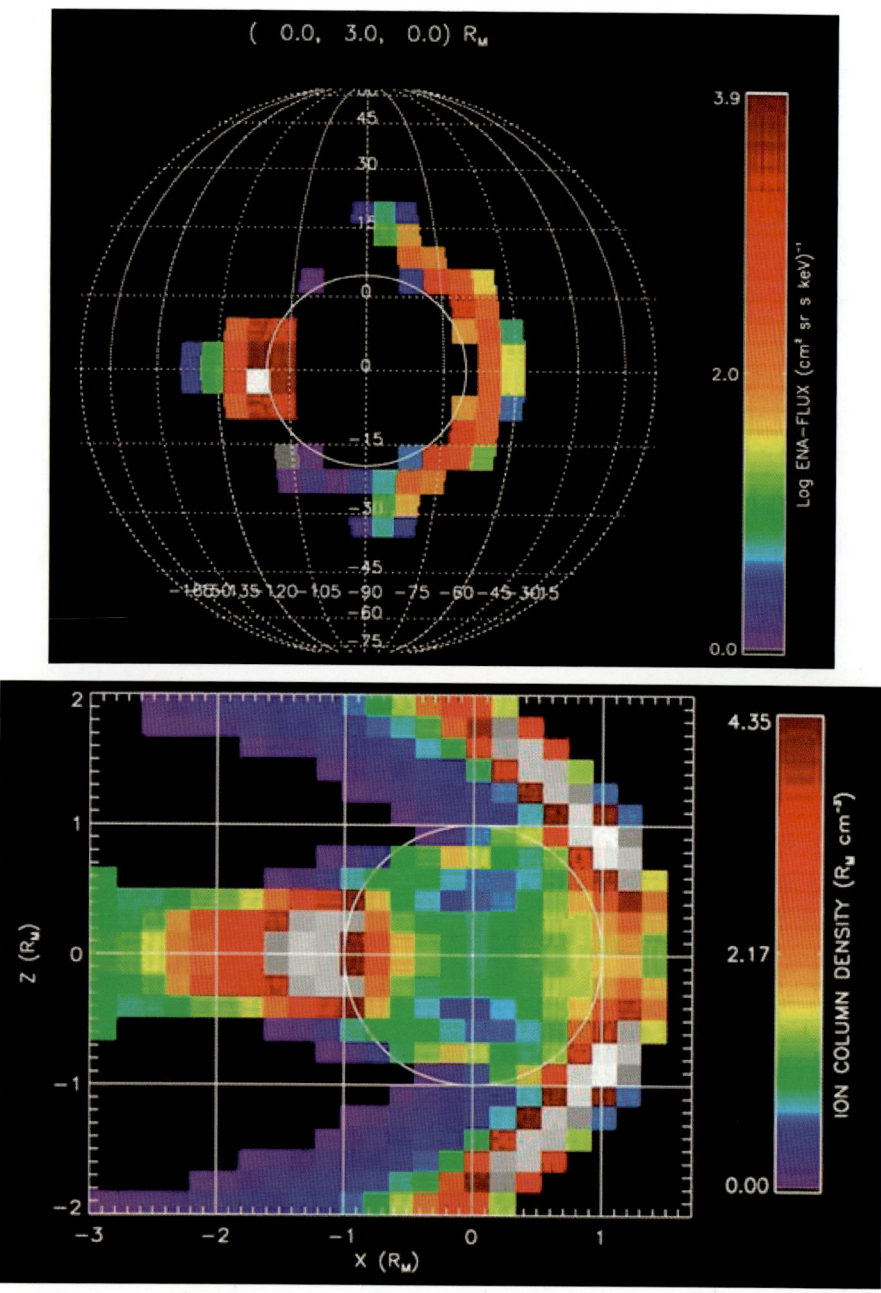

Figure 6-23.

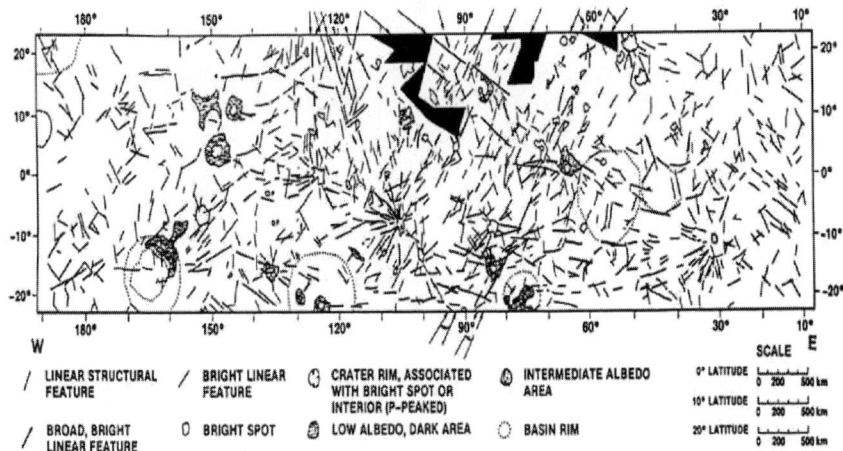

Figure 4-22. Linear Albedo and Structural Features observed in photomosaic (Figure 1.a). Note orthogonal character of bright linear features in the region of highest sun angle (Clark et al, 1988). (Courtesy of Martha Leake.)

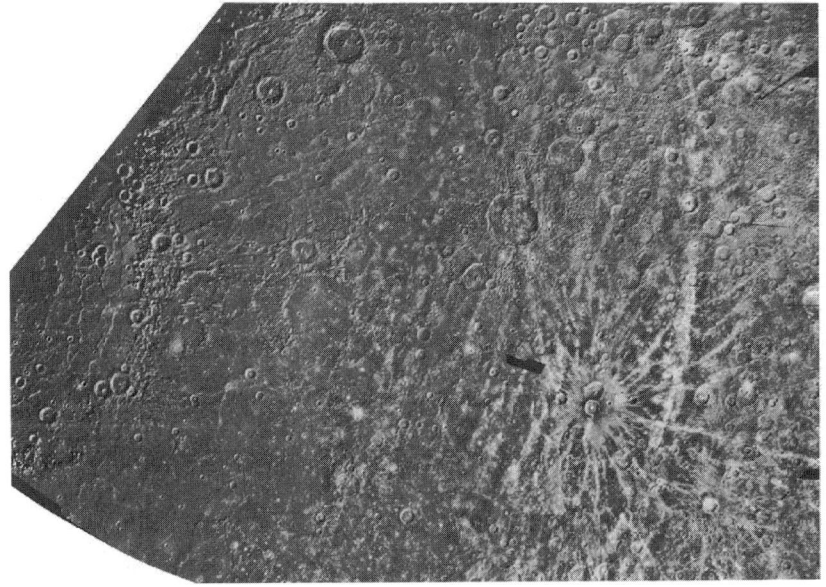

Figure 4-23. Evidence for Tensional Fault Features from NASA Atlas of Mercury. The orthogonal pattern of bright lineaments, in some cases alternating with dark parallel features, is suggestive of highly subdued tensional fault network which should have been associated with crustal expansion during core formation. Some workers have observed evidence of highly subdued topography associated with these features from the limited stereo imaging available (Clark et al, 1988).

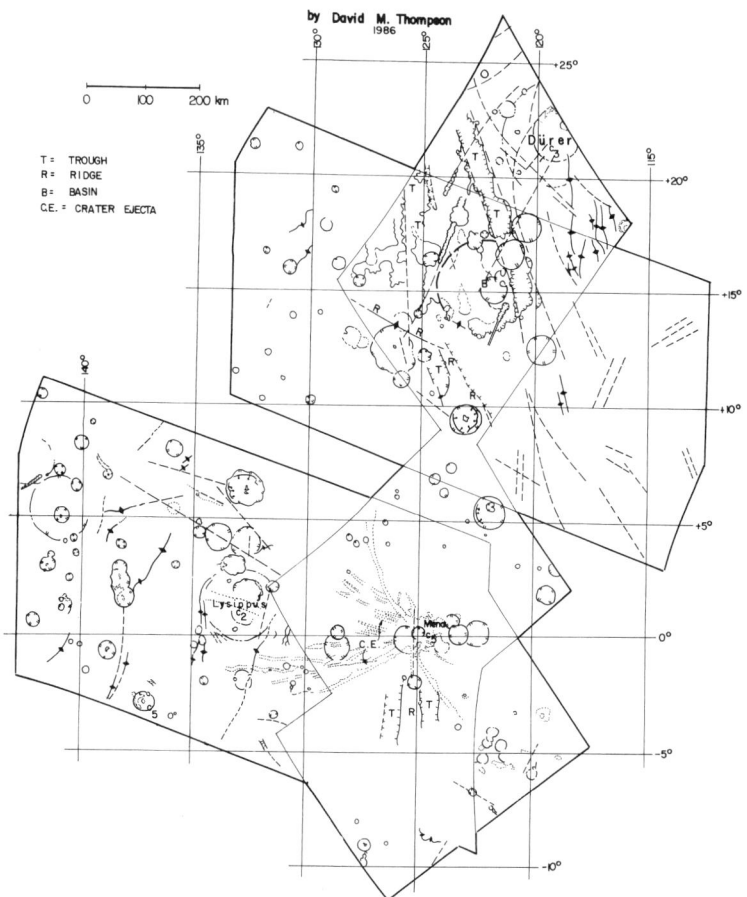

by David M. Thompson
1986

0   100   200 km

T = TROUGH
R = RIDGE
B = BASIN
C.E. = CRATER EJECTA

Figure 4-24. Evidence of Orthogonal Relief Features in Morphological map of portion of H-7 Quadrangle. Associated relief based on Mariner 10 stereo image analysis (From Clark et al, 1988, Goldstone Radar Observations of Mercury, in Mercury, Copyright 1988, The Arizona Board of Regents. Reprinted by permission of the University of Arizona Press.)

tensional features is lacking; however, the plains themselves would have covered up most of the evidence of this fracture system as the eruption process occurred.

Unique to Mercury are the numerous scarps which formed after the intercrater plains (Dzurisin, 1978; Davies et al, 1978). These scarps are compressional features, reverse thrust faults resulting from the crustal contraction which occurred after cooling of the planet's mantle and partial solidification of the core.

A 1 to 2 km decrease in Mercury's radius can be verified on the basis of the observed scarps (Dzurisin, 1978; Strom et al., 1975; Solomon, 1976). Thermal history models predict a two to three times greater decrease in

Figure 4-25. Discovery Scarp from NASA Atlas of Mercury. The most common tectonic feature observed on Mercury, scarps are compressional features, thrust faults, which formed during an episode of crustal shrinking resulting from core consolidation.

radius. Either a much more extensive network of scarps covers the unimaged hemisphere, or the decrease has taken the form of subdued long-wavelength folding not observed due to limitations in the Mariner 10 experiment package (Hauck et al, 2001).

Three types of scarps observed vary in length, height, and degree of linearity. Lobate scarps are linear with regularly scalloped edges, 20 to 500 km long and up to 2 km high with rounded crests. Arcuate scarps are curvilinear, 100 to 600 km long and hundreds of meters high. Irregular scarps are found in crater interiors, less than 100 km long and a couple hundred meters high. These scarps are have steep faces and gently sloping backs, are sinuous and sometimes braided. They are relatively young, transecting most surface features (**Figure 4-25**). Scarps are distributed throughout the observed surface, although the lobate scarps are more prominent in the equatorial region.

Recent studies of large lobate scarps have given insight into scarp formation and the associated properties of the lithosphere (Watters et al, 2001, 2002). Based on a highly effective boundary dislocation model (Watters et al, 2002), the largest observed lobate scarp overlies a 35 to 40 km thrust fault, with a 30 to 35 degree dip, and a 2 km displacement, indicating that large scarps cut through the active lithosphere. Discovery, Adventure, and Resolution Rupes, three large lobate scarps, appear to form a continuous generally northeast trending arc (Watters et al, 2002). Such a distribution is not random and these workers propose that it may be associated with an underlying impact basin structure. Of course scarps, or even impact basin structures, may be reactivating portions of the earliest orthogonal fracture system. A number of workers (Leake et al, 1988; Thomas, 1997) have demonstrated the existence of a preferred distribution of scarps and ridges which shows strong northeast northwest orthogonal trends corresponding to the early Mercury grid produced by tidal despinning.

4.10   POLAR FEATURES

The cold trapping of volatiles in Mercury's polar regions had certainly been proposed (Thomas, 1974), but no convincing evidence of such a phenomenon was observed until recently. Radar mapping of Mercury has revealed anomalous regions (**Figure 4-26**) at the north pole and then the south pole (Slade et al, 1992; Harmon and Slade, 1992; Harmon et al, 2001).

High polarization ratios (ratio of 'depolarized' signal reflected at the same polarization dominated by multiple scatter from <1 cm scale facets 'polarized' signal reflected with the opposite polarization dominated by specular reflection from >10m scale facets) are associated with the highest reflectivities observed for the planet at 12.5 cm and 3.5 cm respectively. Similarly high polarization ratios correlated with high reflectivity at the Martian poles and on icy satellites is diagnostic for the presence of a very cold, clean ice layer. Ironically, no such discrete ice layers have been unambiguously observed for the lunar poles (Nozette et al, 1996; Stacy et al, 1997), which are even colder than Mercury's poles, although ice dispersed in the regolith in the lunar polar regions has probably been detected (Feldman et al, 1998).

High reflectivities are thought to result from the coherent backscatter which occurs when the penetration of the coherent radar signal into the relatively lossless cold ice is followed by multiple scattering due to the fine structure (cracks) in the ice. Radar bright features at the poles were associated with craters as small in diameter as 10 km and as large in diameter as 155 km that were deep enough to have permanently shadowed bottoms which could act as cold traps. The features are the same size at both 12.5 cm and 3.5 cm wavelengths, indicating that a discrete ice layer many wavelengths thick is present. Radar at these wavelengths is thought to penetrate to a depth of approximately 1 meter. The somewhat lower reflectivity for these features than for their Martian analogues are thought to

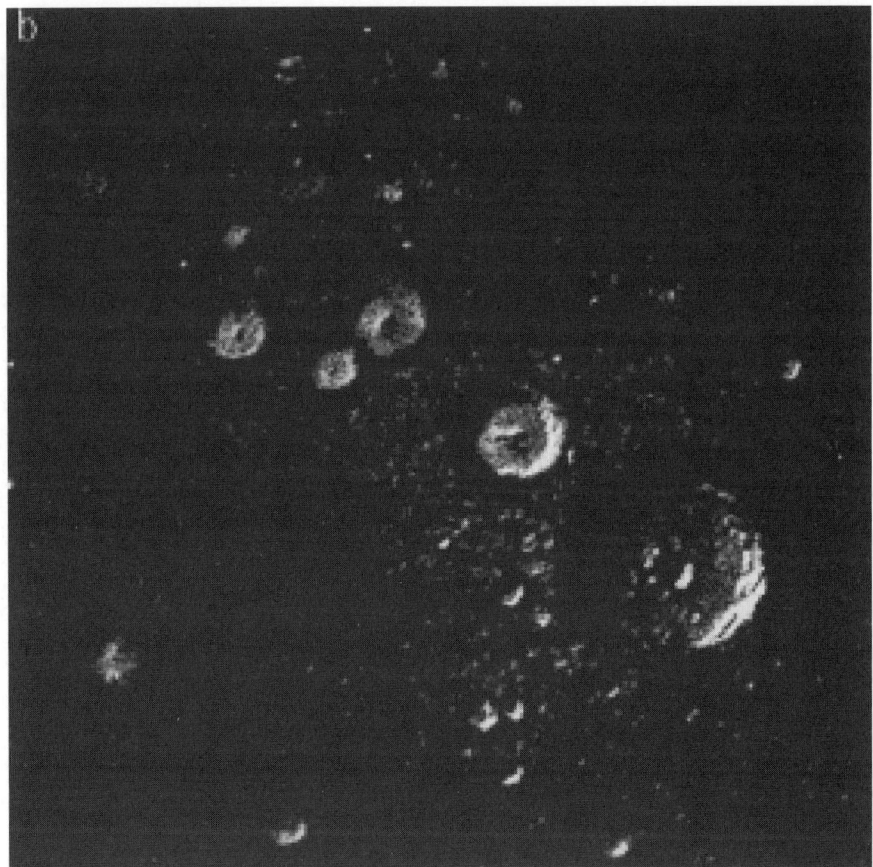

Figure 4-26. Radar detection of polar features. These polar radar reflectivity images from Harmon et al (2001) (taken July 25/26, 1999@resolution 1.5 km) indicate the presence of bright features with characteristics similar to those of Galilean icy satellites at the poles of Mercury. Later work confirmed their association with craters deep enough to retain an ice layer if covered with tens of centimeters of regolith.

result from an insulating dust layer of up to half a meter in thickness over an ice layer of more than half a meter in thickness (Butler et al, 1993; Vasavada et al, 1999).

Numerous models for stability of polar frosts and migration of volatiles in the regolith (e.g. Paige et al., 1992; Ingersoll, 1992; Butler et al, 1993; Salvail and Fanale, 1994; Killen et al., 1997; Feldman et al., 1997; Vasavada et al, 1999) predict that water ice would be marginally stable in permanently shadowed polar or near polar regions over geological time. The most comprehensive model (Vasavada et al, 1999) uses the best available information on crater morphologies (Harmon et al, 1994) and considers the regolith in two layers, based on the best available thermal models (Mitchell and de Pater, 1994): a thin (centimeters scale), loose, insulating layer on a thick (meters scale), consolidated conducting layer. If covered with an insulating layer of dust (Harmon et al, 2001), such layers could exist as far south as 70 degrees latitude. Without such a dust layer, water ice stability would be problematic according to thermal models.

What do the models (Butler et al, 1993; Vasavada et al, 1999) predict are the sources and sinks and the resulting depth of such a layer? An endogenic source for such material would be outgassing from Mercury itself. Such a source is considered unlikely, because of the apparent lack of volcanic activity at the time of formation of the volatile-bearing craters. The more likely exogenic source of volatiles would be impacting meteoritic or cometary impact flux. Cometary impacts are proposed to be more common on Mercury's surface (Strom, 1997) than on other terrestrial planetary surfaces. The extremely pure nature of the volatile is an indication that it was deposited rapidly and migrated to cold traps, in a discrete episode of asteroid or comet impact, tending to support an exogenic volatile rich source, such as Jupiter-family comets (Moses et al, 1999). The lack of evidence for modification of impact structures due to the presence of an endogenous ice sheet supports the model of episodic, exogenic, meteoritic source (Barlow et al, 1999). Dust accumulating on the ice layer after its deposition would insulate it and protect it from further melting. The sinks would include loss due to sputtering at the surface and dissociation in the atmosphere. Developed using conservative estimates for sources and sinks, models (Butler et al, 1993) predict that a net gain in ice accumulation is possible. Even if loss and gain are completely balanced, ice depth could be up to 5 meters if all source water was trapped in polar regions. These models also predict that between 10 and 15 % of the water molecules will become permanently trapped in polar regions.

Cold-trapped sodium (Killen et al, 1990) or sulfur (Sprague et al, 1995, 1996) bearing materials such as iron sulfides have been proposed as alternative polar deposit materials which have greater stability than water at high latitudes over geological time. Killen and coworkers (1990), while observing the sodium exosphere of Mercury, identified enhancements which may have resulted from sodium ions trapped over the polar regions. Implantation and net deposition of such trapped ions in the polar regolith could result. Sodium and potassium enhancements have been associated with large, relatively recent, basin-like features at other latitudes (Sprague, 1992; Sprague et al, 1998) as well, implying diffusion and outgassing from overturned alkali-bearing regolith. These features include Caloris Basin and the antipodal terrain, and the two large radar bright regions, one in northern and the other in southern mid-latitudes, discovered near 340 degrees. Ground-based mid-infrared observations may indicate the presence of alkali bearing materials (Tyler et al, 1988; Sprague et al, 1994, 1998) predicted by models of formation and differentiation of Mercury (Jeanloz et al, 1995). What makes sulfur a potential candidate for the polar volatile deposits?

Those solar system formation models which lead to chondritic or achondritic composition for Mercury predict enhanced abundance of iron sulfides (Wasson, 1988; Cameron et al, 1988). In addition, Sulfur bearing compounds, such as iron sulfide, have high radar transparency and low electrical conductivity. On the other hand, it is less likely that volatiles other than water would be deposited episodically from exogenic sources, creating the thick, clean deposits apparently observed.

Although models predict that Mercury and the Moon should have approximately the same abundance of cold trapped volatiles, Mercury appears to have substantial ice layers, and effectively 'filled' cold traps, whereas the Moon, which is further from the sun and thus colder, gives evidence of less extensive ice deposits (Feldman et al, 1998). Explanations (Vasavada et al, 1999) include the greater obliquity, and thus reduced permanently shaded area, in the Moon's early history, the Moon's lower gravity which would result in less volatile retention, and the greater loss of volatile surface material through sputtering by energetic particles as the Moon orbits through the Earth's magnetospheric tail.

4.11   SUMMARY

Our understanding of Mercury's surface is limited by the resolution and coverage of available images and spectra. Mariner 10 provided the highest resolution coverage, but even those visual images have no better than 100 meter resolution and are only available for one hemisphere. Mariner 10 images provided the basis for our understanding on stratigraphy and surface history resulting from internal volcano-tectonic and external impact processes. Finally, over the last several years, dramatically improved ground-based observational capabilities have resulted in low resolution coverage of the unimaged hemisphere. Ground-based radar observations have provided information on surface morphology all along, and on the presence of buried ice sheets at the poles in recent years. The finding of extensive volatile deposits at Mercury's poles has major implications for surface/exosphere interaction, or space weathering. Models for space weathering have been developed from the study of lunar regolith samples. Mercury is covered with a regolith similar to the Moon's. Evidence for space weathering can be seen in the darkening of Mercury surface deposits in a manner analogous to the darkening which has been observed and studied in collected samples from the Moon.

4.12   REFERENCES

Anderson, J. D., R. F. Jurgens, E. L. Lau, M. A. Slade III, and G. Schubert, Shape and orientation of Mercury from radar ranging data, *Icarus*, 124, 690- 697, 1996.

Barlow, N.G., R.A. Allen, F. Vilas, Mercurian impact craters: Implications for polar ground ice, *Icarus*, **141**, 192-204, 1999.

Blewett, D.T., B.R. Hawke, P.G. Lucey, Lunar pure anorthosite as a spectral analog for Mercury, *Met Plan Sci*, **37**, 1245-1254, 2002.

Blewett, D.T., P. Lucey, B.R. Hawke, G. G. Ling, M.S. Robinson, A comparison of Mercurian reflectance and spectral quantities with those of the Moon, *Icarus*, **129**, 217-231, 1997.

Butler, B. J., and D. O. Muhleman, Mercury radar imaging: Evidence for polar ice, *Science* **258**, 635-640, 1992.

Butler, B.J., D.O. Muhleman, M.A. Slade, Mercury: Full-disk Radar images and the detection and stability of ice at the north pole, *JGR*, **98**, E8, 15003-15023, 1993.

Cameron, A., B. Fegley, W. Benz, and W. Slattery, The strange density of Mercury: Theoretical considerations. In *Mercury,* Vilas, Chapman, Matthews, Eds., U. Arizona, 692-708, 1988.

Chapman, C.R., 1976, Chronology of terrestrial planet evolution: the evidence from Mercury, *Icarus*, **28**, 523-536, 1976.

Chase, S.C., E.D. Miner, D. Morrison, G. Munsch, G. Neugebauer, Mariner 10 infrared radiometer results: temperatures and thermal properties of the surface of Mercury, *Icarus*, 28, 565-578, 1976.

Cintala, M., The Mercurian regolith: An evaluation of impact glass production by micrometeoroid impact. *Lun. Plan. Sci.* **12**, 141-143, 1981.

Cintala, M.J., Impact-induced thermal effects in the lunar and Mercurian regoliths, *JGR*, **97**, E1, 947-973, 1992.

Clark, P. and I. Adler, Utilization of independent solar flux measurements to eliminate nongeochemical variation in X-ray fluorescence data, *Proc.Lunar Planet.Sci.Conf.9th,* 3, 3029-3036, 1978.

Clark, P., S. Joerg, and R. DeHon, 1964, Searching the Sinus Amoris: Using geochemical profiles of geological units, impact and volcanic features to characterize a major terrane interface on the Moon, *Earth, Moon, Plan*, 64, 165-185, 1994.

Clark, P., M. Leake, R. Jurgens, Goldstone radar observations of Mercury. In *Mercury*, F. Vilas, C. Chapman, M. Matthews, Eds., U. Arizona Press, 77-100, 1988.

Clark, P. and J. Trombka, Remote X-ray spectrometry for NEAR and future missions: Modeling and analyzing X-ray production from source to surface, *JGR Planets*, 16361-384, 1997.

Clark, P. and McFadden, L., New results and implications for lunar crustal iron abundance using sensor data diffusion techniques, *JGR Planets,* **105**, E2, 4291-4316, *2000.*

Cooper, B.L., A. Potter, 2004, Mid infrared spectra of Mercury, *JGR Planets* **106** (E12): 32803-32814, 2004.

Cooper, B.L., J.W. Salisbury, R.M. Killen, A.E. Potter, Mid-infrared spectral features of rocks and their powders, *JGR Planets* **107** (E4) #517, 2004.

Crumpler, L. and J. Revenaugh, Hot spots on Earth, Venus, and Mars: Spherical harmonic spectra, *Lun Plan Sci* XXVII, 275-276, 1997.

Davies, M., The control net of Mercury. *RAND R-1914-NASA*, 1-20, 1976.

Dzurisin, D., The tectonic and volcanic history of Mercury as inferred from studies of scarps, ridges, troughs, and other lineaments, *JGR*, 83, 4883-4906, 1978.

Feldman, W.C., B.L. Barraclough, C.J. Hansen, and A.L. Sprague, The neutron signature of Mercury's volatile polar deposits, *JGR* 102, 25656-25574, 1997.

Gaffey, S. and T.B. McCord,, Asteroid Surface Materials: Mineralogical characterizations from reflectance spectra, *Space Science Rev.*, 21, 555-628, 1978.

Gaffey, S., L.A. McFadden, D. Nash, and C.M. Pieters, Ultraviolet, visible, and reflectance spectroscopy: Laboratory spectra of geologic materials. *In Remote Geochemical Analysis: Elemental and Mineralogical Composition*, C.M. Pieters and P.A. Englert, Eds., Cambridge U. Press, 43-77, 1993.

Gold, R., Solomon, S., McNutt, R., Santo, A., Abshire, J., B., Acuna, M., Afzal, R., Anderson, B., Anderson, G., Andrews, P., Bedini, D.,, Cain, J., Cheng, A., Evans, L., Feldman, W., Follas, R., Gloeckler, G., Goldstein, J., Hawkins, E., Izenberg, N., Jaskulek, S., Ketchum, E., Lankton, M., Lohr, D., Mauk, B., McClintock, W. Murchie, S., Schlemm, C., Smith, D., Starr, R. and Zurbucher, T., The MESSENGER mission to Mercury: scientific payload, *Planet. Space Sci.*, **49**, 467-479, 2001.

Grande, M., Investigation of magnetospheric interactions with the Hermean surface, *Adv. Space Res.* **19** (10) 1609- 1614,1997.

Grard, R., Photoemission on the surface of Mercury and related electrical phenomena, *Planet. Space Sci*, 45, 67-72, 1997.

Grard, R., The Coupling of the Mercury Surface and Magnetosphere, http://solar system.estec.esa.nl/planetary/ mercury_ coupling .htm, 1999.

Guest, J. and B. Murray, Volcanic features of the nearside equatorial lunar mare, *J. Geol Soc*, **132**, 251-258, 1976.

Harmon, J.K., Mercury radar studies and lunar comparisons, *Adv Space Res*, **19**, 10, 1487-1496, 1997.

Harmon, J.K., P.J. Perillat, M.A. Slade, High-resolution radar imaging of Mercury's north pole, *Icarus*, **149**, 1-15, 2001.

Harmon, J. and M. Slade Radar mapping of Mercury--Full disk images and polar anomalies, *Science*, **258**, 640- 643, 1992.

Harmon, J. K., M. A. Slade, R. A. Velez, A. Crespo, M. A. Dryer, and J. M. Johnson, Radar mapping of mercury's polar anomalies, *Nature* 369, 212-215, 1994.

Hauck, S.A., A. Dombard, R. Phillips, S. Solomon, Mercury's thermal, tectonic, and magmatic evolution, Mercury: Space environment, surface, and interior, *LPI*, #8004.pdf, 2001.

Head, J., Mode of occurrence and style of emplacement of lunar mare deposits, In Origin of Mare Basalts and their Implications for Lunar Exploration, 6th *Lunar Science,* 66-69, 1975.

Head, J. and Wilson, Lunar mare volcanism: Stratigraphy, eruption conditions, and the evolution of secondary crust, *Geochim Cosmochim Acta*, **56**, 2155-2175, 1992.

Henderson, B.G., B.M. Jakosky, Near-surface thermal gradients and mid-IR emission spectra: A new model including scattering and application to real data, *JGR-Planets*, **102** (E3): 6567-6580 MAR 25, 1997.

Ingersoll, A.P., Thomas Svitek and Bruce C. Murray Stability of polar frosts in spherical bowl-shaped craters on the Moon, Mercury, and Mars. *Icarus* 100, 40-47,1992.

Jeanloz, R., and D. Mitchell, A. Sprague, I. De Pater Evidence for a basalt-free surface on Mercury and implications for internal heat. *Science*, 268, 1455-1457, 1995.

Killen, R. M., J. Benkhoff, and T. H. Morgan, Mercury's polar caps and the generation of an OH exosphere. *Icarus* 125, 195-211 1997.

King, T. V. and W.I. Ridley, Relation of the spectrocopic reflectance of olivine to mineral chemistry and some remote sensing implications; *JGR*, **92**, 11457-11469, 1987.

Ksanfomality, L., Mercury: The image of the planet in the 210-285 W longitude range obtained by the short-exposure method, *Solar System Research*, **37**, 6, 469-479, 2003.

Leake, M.A., C. Chapman, S. Weidenschilling, D. Davis, R. Greenberg, The chronology of Mercury's geological and geophysical evolution: the vulcanoid hypothesis, *Icarus*, **71**, 350-375, 1987.

Ledlow, M.J., J.O. Burns, G.R. Gislee, J.H. Zhao, M. Zeilik, D. Baker, Subsurface emissions from Mercury: VLA radio observations at 2 and 6 centimeters, *Astrophys J*, **384**, 640-655, 1992.

Lewis, J. S., Origin and composition of Mercury in *Mercury*, F. Vilas, C. Chapman, and M. Matthews, Eds., Univ. Arizona Press, pp. 651-666, 1988.

Love, S.G. and K. Keil, Recognizing mercurian meteorites, *Meteoritics*, **30**, 269-278, 1995.

McCord, T., and R. Clark, The Mercury Soil: The presence of Fe2+, *JGR*, **84**, :7664-7668.

Meierhenrich, U.J., W.H. Thiemann, B.Barbier, A. Brack, C. Alzaraz, L. Nahon, R. Wolstencroft, Circular polarization of light by planet mercury and enantiomorphism of its surface minerals, in *Origins of life and evolution of the biosphere*, **32**, 181-190, 2002.

Melosh, J. and McKinnon, W., The tectonics of Mercury. In *Mercury*, U. Arizona, 374-400. 1988.

Mendillo et al, 2001 Imaging the Surface of Mercury, *Plan and Space Science*, **49**, 14-15,

Mitchell, D.L. and I. dePater, Microwave imaging of Mercury's thermal emission at wavelengths from 0.3 to 20.5 cm, *Icarus*, **110**, 2-32, 1994.

Moses, J.I., K. Rawlins, K. Azahnle, L. Dones, External sources of water for Mercury's putative ice deposits, *Icarus*, **137**, 197-221, 1999.

Neukum, G., B.A. Ivanov, R. Wagner, Crater production function and cratering chronology for Mercury, Mercury: Space environment, surface, and interior, *LPI*, #8049.pdf, 2001.

Neukum, G., J. Oberst, H. Hoffmann, R. Wagner, B.A. Ivanov, Geologic evolution and cratering history of Mercury, *Plan and Space Sci*, **49**, 1507-1521, 2001.

Paige, D.A., S.E. Wood, and A.R. Vasavada. . The Thermal Stability of Water Ice at the poles of Mercury. *Science* 258, 643-646. 1992

Pieters, C.M., *Compositional diversity and stratigraphy of the lunar crust derived from reflectance spectroscopy,* In *Remote Geochemical Analyses: Elemental and Mineralogical Composition*, Cambridge University Press, 309-342, 1993.

Potts, L.V., R. von Frese, C.K. Shum, Crustal properties of Mercury by morphometric analysis of multi-ring basins on the Moon and Mars, *Met Plan Sci*, **37**, 1197-1207, 2002.

Reuter, D., D. Jennings, G. McCabe, J. Travis, V. Bly, A. La, T. Nguyen, M. Jhabvala, P. She, and R. Endres, Hyperspectral sensing using the linear Etalong imaging spectral array (LEISA), *In Advanced and Next-Generation Satellites II, Proc Soc Photo-Optical Instr Eng*, Washington, 154-161, 1997.

Robinson, M. and P. Lucey, Recalibrated Mariner 10 color mosaics: Implications for Mercurian volcanism, *Science*, 275, 197-200, 1997.

Robinson, M.S., B.R. Hawke, Low albedo, blue, and opaque rich spectral anomalies in the Mercurian crust, Mercury: Space environment, surface, and interior, *LPI*, #8071.pdf, 2001.

Robinson, M.S., G.J. Taylor, P.G. Lucey, and B.R. Hawke, Complexity of the Mercurian crust, Mercury: Space environment, surface, and interior, *LPI*, #8076.pdf, 2001.

Salvail and Fanale Near-Surface Ice on Mercury and the Moon: A Topographic Thermal Model. *Icarus* 111, 441- 455. 1994.

Schultz, P., Cratering on Mercury: A relook. In *Mercury*, F.Vilas, C. Chapman, M. Matthews Eds., U. Arizona Press, 274-335, 1988.

Slade, M. B. Butler, D. Muhleman Mercury radar imaging: Evidence of polar ice. *Science* 258, 635-640. 1992.

Smith, E.I., Comparison of the crater morphology-size relationship for Mars, Moon, and Mercury, *Icarus*, **28**, 543-550, 1976.

Solomon, S., McNutt, R., Gold, R., Acuna, M., Baker, D., Boynton, W., Chapman, C., Cheng, A., Gloeckler, G., Head, J., Krimigis, S., M., McClintock, W., Murchie, S., Peale, S., Phillips, R., Robinson, M., Slavin, J., Smith, D., Strom, R., Trombka, J., Zuber, M., The MESSENGER Mission to Mercury: scientific objectives and implementation, *Planet. Space Sci.* **49**, 1445-1465, 2001.

Solomon, S., Some aspects of core formation in Mercury, *Icarus*, 28, 509-521, 1976.

Sprague, A. Mercury's atmospheric bright spots and potassium variations: a possible cause, *JGR*, 97, E11, 18257- 18264, 1992.

Sprague, A.L., J.P. Emery, K.L. Donaldson, R.W. Russell, D.K. Lynch, A.L. Mazuk, Mercury: Mid-infrared (3-13.5um) observations show heterogeneous composition, presence of intermediate and basic soil types, and pyroxene, *Met Plan Sci*, **37,** 1255-1268, 2002.

Sprague, A., R. Kozlowski, D. Hunten, F. Grosse An upper limit on neutral calcium in Mercury's atmosphere. *Icarus*, **104**, 33-37, 1993.

Sprague, A., R. Kozlowski, F. Witteborn, D. Cruikshank, D. Wooden, Mercury: Evidence for anorthosite and Basalt from mid-infrared (7.3-13.5 um) spectroscopy, *Icarus,* **109**, 156-167, 1994.

Sprague, A., D. Hunten, K. Lodders Sulfur at Mercury, elemental at the poles and sulfides in the regolith. *Icarus*, **118**, 211-215, 1995.

Sprague, A., D. Hunten, K. Lodders Erratum for Mercury, elemental at the poles and sulfides in the regolith. *Icarus* **123**, 247, 1996.

Sprague, A., D. Hunten, F. Grosse Upper limit for lithium in Mercury's atmosphere. *Icarus,* **123**, 345-349, 1996. 34

Sprague, A., D. Nash, F. Witteborn, D. Cruikshank Mercury's feldspar connection: Mid-IR measurements suggest plagioclase. *Adv. Space Res.* **19** (10), 1507-1510, 1997.

Sprague, A.L. and T.L. Roush, Comparison of laboratory emission spectra with Mercury telescopic data, *Icarus*, **133,** 174-183, 1998.

Sprague, A.L., W.J. Schmitt, R.E. Hill, Mercury: Sodium atmospheric enhancements, radar-bright spots, and visible surface features, *Icarus*, **136**, 60-68, 1998.

Spudis, P., and J. Guest, Stratigraphy and geologic history of Mercury. In *Mercury*, F. Vilas, C. Chapman, and M. Matthews, Eds., U. Arizona Press, 118-164, 1988.

Strom, R.G., *Mercury: The Elusive Planet*, Solar System Series. Smithsonian Institution Press, Washington DC, 1987.

Strom, R., and G. Neukum, The cratering record on Mercury and the origin of impacting objects. In *Mercury*, F. Vilas, C. Chapman, and, M. Matthews, Eds., U. Arizona Press, 336-373, 1988.

Strom, R.G., Sprague, A.., *Exploring Mercury: The Iron Planet.* Spring-Praxis, 1997.

Strom, R., N. Trask, J. Guest, Tectonism and volcanism on Mercury, *JGR*, 80, 2478-2507, 1975.

Taylor, S.R., Solar System Evolution (Publ. Cambridge University Press), 460 p., 2001.

Thomas, P., Are there other tectonics than tidal despinning, global contraction and Caloris related events on Mercury? A review of questions and problems, *Plan Space Sci*, **45**, 3-13, 1997.

Trask, N.J. and R.G. Strom, Additional evidence of Mercurian volcanism, *Icarus,* **28**, 559-564, 1976.

Tyler, A.L., R. Kozlowski, L.A. Lebofsky, Determination of rock type on Mercury and the Moon through remote sensing in the thermal infrared, *GRL*, **15**, 8808-811, 1988.

Uchupi, E. and K. Emery, Mercury, in *Morphology of the rock members of the solar system*, 121-136, 2003.

Vasavada, A.R., D.A. Paige, S.E. Wood, Near-surface temperatures on Mercury and the Moon and the stability of polar ice deposits*, Icarus*, **141**, 179-193, 1999.

Vilas, F. Mercury: Absence of crystalline Fe+2 in the regolith. *Icarus*, **64**, 133-138, 1985.

Vilas, F. Surface composition of Mercury from reflectance spectrophotometry. In *Mercury*, F. Vilas, C. Chapman, M. Matthews, Eds., Univ. Arizona Press, 59-76, 1988.

Vilas, F., M.A. Leake, W. Mendell, The dependence of reflectance spectra of Mercury on surface terrain, *Icarus*, **59**, 60-68, 1984.

Wagner, R.J., U. Wolf, B.A. Ivanov, G. Neukum, Application of an updated impact cratering chronology model to Mercury's time-stratigraphic system, Mercury: Space environment, surface, and interior, *LPI*, #8049.pdf, 2001.

Wanke H., Constitution of terrestrial planets; *Phil. Trans. R. Soc.Lond.* A 303, p. 287-302, 1981.

Warell, J. and S.S. Limaye, Properties of the Hermean regolith: I. Global regolith albedo variation at 200 km scale from multicolor CCD imaging, *Plan Spcae Sci*, **49**, 1531-1552, 2001.

Warell, J., Properties of the Hermean regolith: II. Disk-resolved multicolor photometry and color variations of the 'unknown' hemisphere*, Icarus*, in press.

Warell, J. Properties of the Hermean regolith: III. Disk-resolved vis-NIR reflectance spectra and implications for the abundance of iron, *Icarus*, in press.

Wasson, J. T., The building stones of the planets, in *Mercury*, F. Vilas, C. Chapman, and M. Matthews, Eds., Univ. Arizona Press, pp. 622-650, 1988.

Watters, T.R., A.C. Cook, M.S. Robinson, Large-scale lobate scarps in the southern hemisphere of Mercury, *Plan Space Sci*, **49**, 1523-1530, 2001.

Watters, T.R., R.A. Schultz, M.S. Robinson, A.C. Cook, The mechanical and thermal structure of Mercury's early lithosphere, *GRL*,**29**, 11, 10.1029/2001GL014308, 37, 2002.

Wetherill, G., 1988. Accumulation of Mercury from planetesimals. In *Mercury*, F. Vilas, C. Chapman, and M. Matthews, Eds., U. Arizona, 670-691.

Whitford-Stark, J. and J. Head, The Procellarum volcanic complexes: Contrasting styles of volcanism, *Proc Lun Sci Conf 8th*, 2705-2724, 1977.

Wilhelms, D., Mercurian volcanism questioned, *Icarus*, **28**, 551-558, 1976.

Wilhelms, D., The geologic history of the Moon, *U.S. Geol.Surv.Prof. Paper* 1348, 1987.

4.13   SOME QUESTIONS FOR DISCUSSION

1. How can the presence of volatiles at Mercury's poles be explained?

2. Describe the nature, resolution, and coverage of data required to resolve questions about the origin of Mercury's major terranes.

3. Discuss the role of volcanism in the formation of Mercury's surface based on what we know now.

4. Speculate on what we will find in the unimaged hemisphere.

Chapter 5

MERCURY'S EXOSPHERE

5.1 THE EXOSPHERE CONCEPT

An exosphere is an ensemble of atoms or molecules above a planet's surface or atmosphere for which the mean free path is greater than the scale height (the e-folding height for density). Because collisions are rare in an exosphere, each constituent maintains its own distribution, defined by its unique combination of source energy and distribution, mass, radiation pressure and loss processes (e.g. Jeans escape, photo-dissociation; adsorption; ionization; surface chemistry). Killen and Ip (1999) give a good discussion of this concept, and many additional references.

5.2 FROM ATMOSPHERE TO EXOSPHERE

In Zolner's book on Mercury published in 1874, it was noted that this body "probably does not hold a noticeable atmosphere". On the other hand, Schiaparelli thought that Mercury's surface features appeared to be more pronounced at some times than at others, suggesting that they were being viewed through a haze. On the basis of this erroneous assumption, he speculated in 1899 that one permanently heated Hermean hemisphere adjacent to a perpetually frozen one caused strong atmospheric circulation, thereby producing a rather even atmospheric temperature. Determinations of surface temperatures on Mercury which were not available in Schiaparelli's day show them to range between 90 K and 740 K, the most extreme variation found on any Solar System body.

If it is assumed that Mercury had an atmosphere like that of the Earth then, on its sunward side, the hydrogen component would quickly escape. Also, oxygen would rapidly disappear through forming compounds both with exposed, un-oxidized, metals and with oxidizable minerals. As a result of flow to the dark side of the planet, all atmospheric constituents unable to remain in a gaseous state at the darkside temperature would change phase and precipitate out. Calculation shows that, while this depletion of atmospheric species is somewhat slowed because of the formation (due to nutation) of a 'twilight belt' at the terminator where fluctuating temperature conditions prevail, this process could be completed in two centuries or less, leaving the planet with a very sparse atmosphere. See also arguments concerning the sparse nature of the Mercurian atmosphere presented by Eugene Antoniadi in 1934 in the second book ever to be published on the planet Mercury.

In 1974-75, during the Mariner 10 flybys, in situ atmospheric data were obtained directly at the planet. From these measurements an upper limit for the gas density on the dayside surface ($\sim 10^6$ cm^{-3}) was derived and it was established, in conformity with a prediction of Banks et al (1970), that Mercury's atmosphere is exospheric down to the planet's surface. Traces of hydrogen, helium and oxygen were identified and these constituents were deduced to form individual exospheres. Upper abundance limits for the presence at the planet of neon, argon and carbon were also established.

Prior to the Mariner 10 encounter, Banks et al (1970) reported no observational evidence for the presence of an atmosphere surrounding Mercury, but predicted the presence of an exosphere (of column abundance $N < 2 \times 10^{14}$ cm^{-2}) resulting from diffusion or effusion from the interior, as well as from solar wind implantation, with subsequent thermal evaporation or sputtering. It was anticipated that the most abundant constituents would be noble gases (from the solar wind and from radiogenic sources). In descending order of abundance, ^{4}He, ^{40}Ar and ^{20}Ne were predicted to be present. Fink et al. (1974) argued that CO_2 might be frozen out on the dark-side. On the basis of these assumptions, the Mariner 10 ultraviolet spectrometer was designed to search for H, He, O, C, CO, CO_2 and the noble gases argon and neon.

5.3   MARINER 10 OBSERVATIONS

During the three Mariner 10 flybys of Mercury, information concerning the planet's atmosphere was obtained from airglow and occultation measurements made by the onboard ultraviolet spectrometer (UVS) (Broadfoot et al, 1974, 1976; Kumar, 1976). An upper limit for the gas density on the dayside surface (about 10^6 cm^{-3}) was established from these data and it was confirmed that Mercury's atmosphere is exospheric down to the surface. In addition, helium and atomic hydrogen were positively identified, and atomic oxygen tentatively identified, as atmospheric constituents. Upper limits on number densities at the terminator were placed on He, Na, K, O, Ar, H_2, O_2, N_2, CO_2 and H_2O using the radio occultation experiment. These measurements, however, are uncertain due to the nature of the measurement technique. The airglow spectrometer provided column densities, but surface density can be calculated by taking the surface temperature and assuming hydrostatic equilibrium (barometric vertical distribution) to obtain the scale height. Together with the measured column density, this will then give an estimate of surface density. None of the species found so far are in thermal equilibrium with the surface.

The helium and hydrogen atmospheres were measured several thousand kilometers above the surface and, at these altitudes, they displayed velocity distributions that were approximately thermal in character but with different temperatures for each constituent. No noble gases or molecules were found. In the hydrogen detected at the subsolar point there were two thermal components, one characteristic of the day-side temperature (420 K) with surface number density 23 and the second characteristic of the nightside temperature (−110 K) with surface number density 230. Several explanations were advanced for the cold component including: 1) a night-side source (Shemansky and Broadfoot, 1977), 2) pyrolysis of water (Broadfoot et al, 1976), and 3) surface chemistry (Potter, 1995). The presence of the cold component is still not fully understood (Killen and Ip, 1999). Of particular note, the lateral distributions did not follow the expected ($nT^{5/2}$ = constant) law for an exosphere in thermal equilibrium with its related surface temperature.

Early models failed to reproduce the observed distributions of H and He in Mercury's atmosphere (See Killen and Ip, 1999). This failure was attributed to 1) the non-Maxwellian source velocity of proposed source-generating mechanisms, such as photon-stimulated desorption discussed below, and 2) the lack of thermal accommodation with the surface (Shemansky and Broadfoot, 1977).

5.4   POST-MARINER 10 UNDERSTANDING OF MERCURY'S ATMOSPHERE

Mercury's tenuous neutral atmosphere was confirmed by Mariner 10 observations to be a proper exosphere. The exobase is at the surface (i.e., an atmospheric neutral will typically fall back to the surface of the planet before colliding with another neutral). Magnetospheric processes (including ion precipitation onto Mercury's surface and the pickup of photo-ions), were recognized to be extremely important for both atmospheric sources and losses. (Pickup is the process whereby newly created ions are immediately swept away under the convective action of magnetospheric electric and magnetic fields.)

The origin of Mercury's exosphere was variously related in early post Mariner 10 literature (e.g., Shemansky and Broadfoot, 1977) to: 1) Solar Wind accretion, 2) a supply of exospheric components through the radioactive decay of uranium and thorium in the planet's crust, and 3) a large supply of He due to the proximity of the Sun.

Hartle et al. (1975) calculated that only a fraction (6.2×10^{-4}) of the solar wind He^{++} (alpha particle) flux intercepted by the Hermean magnetosphere is required to maintain the observed exosphere, if this flux is fully absorbed by the planet's surface (i.e., if its outermost layers are saturated by helium). Although this was demonstrated to be the case for lunar ^4He and ^{20}Ne, investigators recognized that the rates of processes such as erosion and gardening (regolith overturn) might be different for Mercury. Nevertheless, the flux of neutral helium out of Mercury's surface calculated by this model was comparable to the outgassing rate of He produced by the radioactive decay of thorium and uranium in the planet's interior. This assumes that similar amounts of these two elements had condensed from the solar nebula during the formation of Mercury and the Earth (Lewis, 1972). Only detailed observations of the He airglow around the entire planet would distinguish which is the dominant source.

From proton and He^{++} observations made aboard the Helios spacecraft at 0.3 to 0.4 AU, it was already known that the alpha particle to proton ratio can frequently increase for a few hours from a 'normal' value of about 10% to between 10-50%. As the solar wind accretion at Mercury is larger and faster than at the Earth, any dramatic change in the He^{++} content in the interplanetary medium would imply significant changes in the He airglow intensity. If alpha particles can leak into Mercury's magnetosphere, then this should also be the case for the more abundant solar wind protons (Kallio and Janhunun, 2004; Kabin and Gombosi, 2000), and for C, O and N, which are probably more abundant than Ne.

If Mercury's surface is fully reduced, protons will recombine to form molecular hydrogen (column density 8.5×10^2 cm^{-2}) (Hartle et al, 1973). If the surface is not fully reduced, these protons will be oxidized to form water (3.5×10^3 cm^{-3}) which will photo-dissociate into atomic hydrogen (Hartle et al, 1973). Both components are detectable in Lyman alpha (1216Å) but with an intensity that falls below the all-sky background of the Mariner-10 airglow spectrometer (Hartle et al. 1975).

If Mercury's surface is not fully oxidized, oxygen ions could be absorbed and ions of C and N could also remain implanted in the surface. Otherwise, CO and CO_2 could be formed and, if the surface is fully reduced, even the formation of CH_4 and NH_3 could be possible. If the components mentioned are not adsorbed, or chemisorbed by Mercury's surface, they should, in principle, contribute to the Hermean atmosphere as their ballistic flight times are much smaller than their lifetimes with respect to photo-ionization by solar UV radiation (Kumar, 1976; Huebner et al, 1992; Hartle and Killen, 2006). Even if these species are photo-dissociated and photo-ionized, a solar wind induced magnetospheric convection would transport them to the interplanetary medium where they could, in principle, be detected by a solar wind composition experiment.

Except for H and He, none of the components mentioned above were detected by the Mariner 10 airglow spectrometer (the ion analyzer of the solar wind experiment was not functioning). Also, there was no information concerning Ar isotope ratios (particularly $^{40}Ar/^{36}Ar$) which would allow a critical test to be made of the radio active decay hypothesis of the origin of Mercury's atmosphere.

Thus, even immediately after the Mariner 10 flybys, investigators already recognized the need for a new mission carrying a solar wind composition detector and an improved, low background, UV spectrometer covering the most characteristic wavelength channels of the above mentioned atoms and molecules. They realized that such measurements would not only provide information on the nature and origin of Mercury's atmosphere but also on the state of reduction of the Hermean surface. This would provide an important contribution to understanding key geochemical processes related to the overall evolution of the planet.

Turning next to the Hermean ionosphere, dispersive frequency measurements by the Mariner 10 dual frequency (S- and X-band) radio occultation experiment yielded only an upper limit for the day-side electron density (10^3 cm^{-3}) (Fjeldbo et al. 1976). The Mariner 10 electron experiment observed an electron density of 0.1 cm^{-3} within the polar region and on an open magnetic flux tube. Elsewhere in the magnetosphere the density was 1 cm^{-3}. For an ionosphere composed largely of He^+ and electrons, the electron density would be 1.6×10^{-2} cm^{-3} if the observed atmospheric values for He were used (Ogilvie et al. 1977. Some as yet unanswered questions are: 1) Is there a Hermean ionosphere? 2) If the Mariner 10 observations of are correct, is this ionosphere solar wind induced or of atmospheric origin? 3) What is its ionic composition?

5.5   GROUND BASED OBSERVATIONS OF SODIUM AND POTASSIUM

Well after the Mariner 10 flybys, a fortuitous observation led to the identification of sodium and potassium exospheres at Mercury. These first observations were made using ground based instrumentation consisting originally of single-slit, spectrographs (Potter and Morgan, 1985, 1986). Later, an image slicer technique was introduced (Potter and Morgan, 1990). An account of this latter method is contained in Pierce (1965). (Briefly, an image slicer placed at the entrance to the Echelle spectrograph produces a data cube characterized by two spatial dimensions and one spectral dimension.) Their ground based observations identified Na and K at Mercury associated with strong local planetary sources characterized by short durations (Potter and Morgan, 1985, 1986). At the present time the multi-constituent exosphere is considered to form part of a coupled system with the surface at its base and with the particle, magnetic field and interplanetary environment acting as both a source and sink for neutral atoms. The interaction of the Solar Wind and exosphere with the magnetosphere is still relatively poorly understood. The hope of resolving current uncertainties inspires the design of sophisticated ground-based observations programs as well as experiments to be flown on future missions to the planet.

Data recorded by Potter and Morgan (1985) consists of intense, sodium D-line emissions (5890 Å, 5896 Å) excited by solar resonance scattering in Mercury's atmosphere. **Figure 5-1** presents the relevant spectrum region where Mercury lines are almost hidden by solar absorption. Thereafter, Potter and Morgan (1986) recorded similar, but less bright, potassium emissions (7665 Å, 7699 Å) at Mercury. Typical sodium subsolar column abundance was $1-3 \times 10^{11}$ atoms cm^{-2} (Potter and Morgan, 1985, 1986). Potassium has been found to be about two orders of magnitude less abundant than sodium (Potter et al, 2002), with Na/K ratios averaging around 100 but varying from 40 to 140 (Potter et al, 2002).

Variations in the sodium (Killen et al. 1990) and potassium emission brightnesses have been observed along the observing slit (Potter and Morgan, 1986). The distribution of the alkalis was often found to peak at high to middle latitudes, decreasing towards the terminator. High latitude enhancements were often observed when the planet was imaged in sodium (Potter and Morgan, 1990; Potter et al, 2002). Only the sodium exosphere was thereafter monitored over extended timescales, using the strong resonance transitions in the D-region. These observations revealed planet wide changes in the brightness distributions, often with one or two localized regions waxing or waning on a timescale of days or months (Potter and Morgan, 1990, 1997; Potter et al,1999; Potter et al, 2002). **Figure 5-2** presents a snapshot of representative bright sodium regions on Mercury's surface on October 5, 2003.

It was suggested by Potter and Morgan (1990) that the high latitude variations provide evidence of solar wind-magnetosphere related sputtering of sodium and potassium from the Hermean surface. An alternative recycling mechanism proposed by Sprague et al (1998) explained the variations in terms of the implantation of sodium and potassium into grains

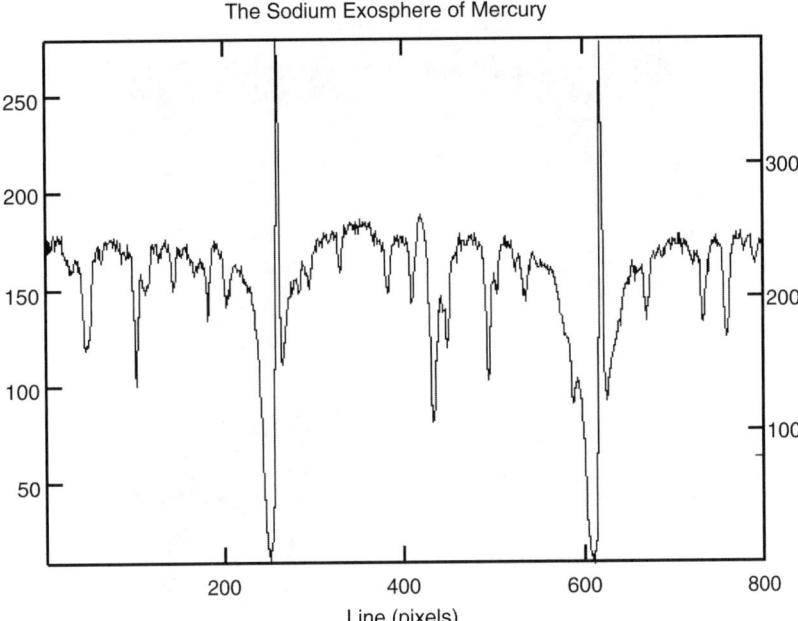

Figure 5-1. Spectrum of Mercury Showing Sodium Lines almost hidden in the solar absorption (from unpublished results of Killen and Potter).

at the regolith surface during the long (88-day) Hermean night, followed by their diffusion to the vacuum of space when the enriched surface of the planet rotated into the sunlight. It was noted in support of the latter interpretation that the variable sodium and potassium signatures display a marked morning/afternoon asymmetry which would be an expected consequence of the time and temperature dependent out-gassing proposed. In related work, Potter (1993) pointed out that time varying bright spots are also distributed near the evening portion of the planet. Because sunrise and sunset hemispheres are never observed simultaneously, temporal and spatial effects are mixed, and certain true anomaly angles have never been observed.

Sprague et al (1990) reported an enhancement in both sodium and potassium was reported to be present in the longitude range containing the Caloris Basin, and they attributed this observation to increased diffusion from a related deep source. Potter and Morgan (1997) pointed out, that the enhancement observed was consistent with signatures simultaneously present at other longitudes not associated with this geological feature. They then proposed, based on modeling, that increased the ion sputtering resulting from ions entering the cusp region provides a more probable explanation for the large increase in the sodium content of the exosphere observed (Killen et al, 2001). Killen et al (2001) showed that a variation in Na content of a factor of 3 was observed in a week long series of observations during which

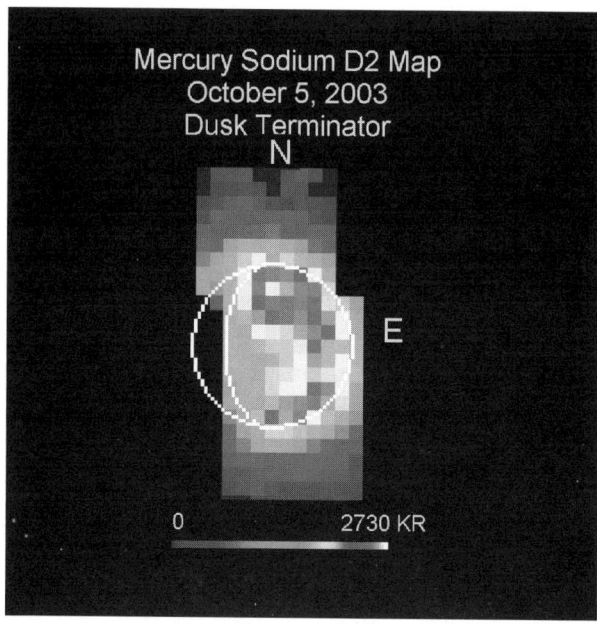

Figure 5-2. Sodium Emission from Mercury. Typical emission showing bright spots from unpublished measurements of Killen and Potter. **See Color Plate 2.**

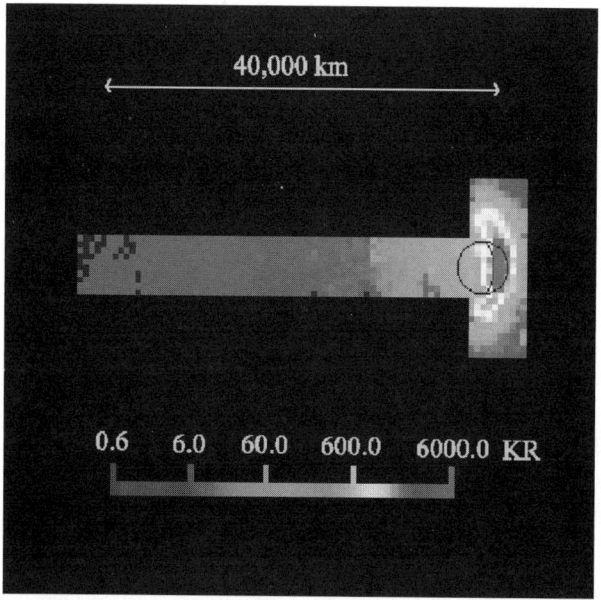

Figure 5-3. The Sodium Tail of Mercury (From Potter, Killen, and Morgan, 2002, reprinted with permission of Meteoritics and Planetary Science, Copyright by the Meteoritical Society). See Color Plate 2.

which the Caloris Basin remained in the field of view with almost no change. These variations could not be attributed to changes in the geology of the region observed or to the morning/evening effects. There is no correlation between column abundance and Hermean longitude in the observations made by Potter et al (2005).

5.6 THE SODIUM TAIL OF MERCURY

Models of the sodium atmosphere of Mercury by Smyth and Marconi (1985) as well as by Ip (1986) predicted the possible existence of a comet like sodium tail at Mercury. Potter et al (2002) used the 1.5m McMath Pierce Solar Telescope at Kitt Peak in Arizona to seek, and successfully find, the sodium tail of Mercury illustrated in **Figure 5-3**. The fact that an extended sodium tail was detected indicates that radiation pressure acceleration is sufficient to accelerate Na to escape velocity as long as the initial velocity range is >2 km/second. It is estimated that about 10% of the Na in the atmosphere escapes down the tail.

Ion sputtering was suggested by McGrath et al. (1986) to provide a likely loss mechanism for sodium atoms. However, this process alone could not account for all of the observed sodium. The bulk of sodium at Mercury displays translational temperatures near 1000 K (Killen et al, 1999) which, as shown by Madey et al (1998), roughly corresponds to the temperature necessary for photon sputtering. It should also be noted that while much of the sodium at Mercury may be produced by photon bombardment, little of this component actually escapes because of its relatively low velocity. A complete model of the sodium exosphere must ultimately include the contributions of ion, photon, and meteoritic sputtering, and the predictions of this model compared with measured data to gain an understanding of the relative importance of the two concerned processes. A model by Killen et al (2001) concluded that 10 to 20% of the atmosphere is produced by meteoritic vaporization, and up to 30% is produced by ion-sputtering. The ion sputtered component is the most variable and is correlated to the interplanetary magnetic field (Sarantos et al, 2001; Kallio and Janhunen, 2004).

5.7 DISCOVERY OF CALCIUM IN MERCURY'S ATMOSPHERE

Sprague et al (1993) measured an upper limit of 7.4×10^6 cm^{-2} for the neutral calcium abundance at Mercury. Killen and Ip (1999) noted that, because Ca is refractory, it constitutes a signature species for ion sputtering.

Calcium was, thereafter, discovered in the atmosphere near Mercury's poles using data recorded by the High Resolution Echelle Spectrograph (HRES) at the W.M. Keck I telescope (Bida et al, 2000). The Ca zenith abundance (1.0 x 10^7 cm^{-2}) was relatively low relative to the value (6.4 x 10^8 cm$^{-2)}$ predicted by Morgan and Killen (1997), while the temperature was, apparently, high (12,000K). The localized distribution and high temperature suggested that the calcium detected was produced by surface sputtering by ions that entered Mercury's auroral zone or by dissociation of an oxide in the atmosphere (Killen et al, 2004). The low density and apparent polar concentration of the calcium detected are consistent with a surface composition that is more volatile rich than that of the Moon (Morgan and Killen, 1997). It is likely that calcium is vaporized in molecular form (Killen et al, 2004).

5.8   MERCURY'S EXOSPHERE AFTER SODIUM AND POTASSIUM DETECTION

Following the ground based discovery of sodium and potassium in the Hermean atmosphere, Hunten et al (1988) produced an excellent review of the state of knowledge concerning Mercury's exosphere. These authors pointed out that, although Mercury's atmosphere is technically an exosphere, the gas-surface interaction is very different from the interaction of an exosphere in contact with an underlying atmosphere.

Mercury's exosphere is the observable result of a dynamically coupled system, with the planetary surface acting as both source and sink, and the magnetosphere as both buffer and accelerator for solar wind plasma. The low density of volatiles from the surface and the small internal magnetic field, combine to produce this dynamic environment, which is termed a surface-bounded exosphere.

The observed atmospheric species distributions are derived from two distinct populations: source atoms extracted directly from the regolith as well as ambient atoms that have undergone at least one ballistic orbit and are periodically adsorbed on the surface (Killen and Ip, 1999). More energetic processes capable of breaking chemical bonds are required to generate source atoms.

Quantum mechanical effects alter the velocity distribution and the rate of migration of ions across the planet's surface (Shemansky, 1988). Also, accommodation to the local temperature is inefficient (Killen et al, 1999). Possible sources of exospheric constituents vary greatly by element. The solar wind has been proposed as the source for hydrogen and helium (Hartle et al,1975). Evaporation of meteoric material provide some of the alkalis and water, with a possible dominant contribution to the former from ion-sputtering and photo-sputtering (Killen et al, 2001). Solar wind ions are not

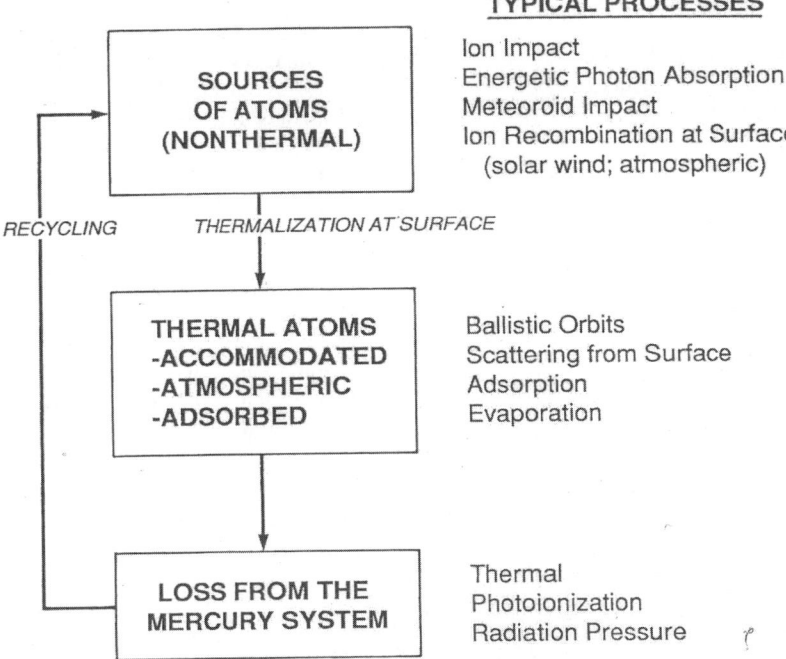

Figure 5-4. Simple Flow Diagram for Mercury's Atmosphere (From Hunten et al, 1988, The Atmosphere of Mercury, in Mercury, Copyright 1988, The Arizona Board of Regents. Reprinted by permission of the University of Arizona Press.)

generally considered to enter the atmosphere directly but, rather, via Mercury's magnetosphere (Kallio and Janhunan, 2004). These may be implanted in surface materials and to be later displaced in the atmosphere due to the impact of subsequently arriving ions. Hydrogen is probably mostly released as H_2 (Hartle, 1973).

Figure 5.4 is a simple flow diagram showing the relationship between sources and sinks for an atmospheric constituent (Hunten et al, 1988). A current review of the atmosphere, which quite significantly updates the information compiled by Hunten et al. (1988) has been provided by Killen and Ip (1999).

Table 5.1 lists measured and predicted abundances of those constituents of Mercury's atmosphere for which data are presently available, with upper limits given for those for which estimates are not available, as determined by Killen and Ip (1999). An upper limit for the abundance of lithium ($<8.4. \times .10^7$ cm^{-2} at the local zenith) (Sprague et al, 1996) is included as well.

There are certainly many more species that have not yet been detected in Mercury's atmosphere. Current understanding of source processes suggests the presence of Ar, Si, Al, Mg, Fe, S, and OH. With the exception of Ar, all these elements have strong ground state emission lines in the range 0.13 to

Table 5-1. Mercury Atmospheric Species Abundances

	Surface Abundance (cm⁻³)	Total Zenith Col (cm⁻²)	λ (A)	Photoion LossRate (cm⁻² s⁻¹)	References for Abundances
H	23-230	3×10^9	1025 1216	1.5×10^3	Hunten et al, 1988; Shemansky, 1988 Mariner 10; Killen and Ip, 1999
He	6.0×10^3	3×10^{11}	584	1.1×10^5	Hunten et al, 1988; Shemansky, 1988 Mariner 10; Killen and Ip, 1999
O	4.4×10^4	3×10^{11}	1304	4.0×10^5	Hunten et al, 1988; Shemansky, 1988 Mariner 10; Killen and Ip, 1999
Na	3.7×10^4	2×10^{11}	3304	9×10^6 to 2×10^7	Hunten et al, 1988; Killen et al, 1990 Ground-based; Killen and Ip, 1999
K	3.2×10^1	1×10^9	4044	1.5×10^5	Potter and Morgan, 1988 Ground-based; Killen and Ip, 1999
Ar	$<6.6\times10^6$	$<2\times10^{13}$	1048	4.2×10^7	Hunten et al, 1988; Shemansky, 1988 Upper limit; Killen and Ip, 1999
^{20}Ne	6.0×10^3 day 7.0×10^5 nite	3.7×10^{10}	736		Hodges, 1974; Shemansky, 1988 Prediction
H₂	$<2.6\times10^7$	$<8.7\times10^{14}$	1608	8.8×10^8	Hunten et al, 1988; Shemansky, 1988 Prediction; Killen and Ip, 1999
O₂	$<2.5\times10^7$	$<9.6\times10^{13}$		2.7×10^9	Hunten et al, 1988; Shemansky, 1988 Prediction; Killen and Ip, 1999
N₂+	$<2.3\times10^7$	$<1\times10^{10}$	3914	4.5×10^4	Hunten et al, 1988; Shemansky, 1988 Prediction; Killen and Ip, 1999
CO₂ +	$<1.6\times10^7$	$<4.5\times10^{13}$	2890	2.0×10^8	Hunten et al, 1988; Shemansky, 1988 Prediction; Killen and Ip, 1999
H₂O	$<1.5\times10^7$	$<1\times10^{12}$ to $<1\times10^{14}$	1.38 mm	2.0×10^8	Hunten et al, 1988; Killen et al, 1990; Shemansky, 1988; Prediction; Killen and Ip, 1999
OH	1.4×10^3	$>1\times10^{10}$	3085	2.7×10^6	Killen et al, 1990; Killen and Ip, 1999 Morgan and Killen, 1997 Prediction
Mg	7.5×10^3	3.9×10^{10}	2851 2801		Killen and Ip, 1999; Morgan and Killen, 1997 Prediction
Ca	$<239-387$	1.0×10^8	4227		Bida et al, 2000; Killen, Bida, Morgan, 2004
Fe	340	7.5×10^8	3719		Morgan and Killen, 1997 Prediction
Si	2.7×10^3	1.2×10^{10}	2526		Morgan and Killen, 1997 Prediction
S	5.0×10^3 to 6.0×10^5	2.0×10^{10} to 2.0×10^{12}	1813	1.5×10^5 to 1.5×10^8	Morgan and Killen, 1997; Sprague et al, 1995 Prediction
Al	654	3.0×10^{19}	3092		Hodges, 1974; Morgan and Killen, 1997 Prediction
Li		8.4×10^7	6708		Sprague et al, 1996 Upper limit

0.43 microns (Morgan and Killen, 1997). However, because their emission lines lie in the UV, these signatures have not yet been observed.

Because Mercury's atmosphere is exospheric in nature, this planet can be considered to possess multi-atmospheres, each of which forms independently and each of which has the capacity to be very different. The differences result from the unique properties of each particular gas, the nature of the sources and sinks for that species, and its interactions with the surrounding environment.

5.9   CURRENT UNDERSTANDING OF SOURCE AND LOSS PROCESSES

The simplest loss process is thermal release (Jeans escape), which is caused by heating of the surface due to solar radiation. Given the large surface temperature on the dayside (Chapter 1), very volatile molecules can evaporate into space with scale heights and column densities corresponding to the local surface temperature. Only H and He escape Mercury predominantly through this mechanism.

Thermal vaporization is too rapid to be sustained at temperatures near the sub-solar point on Mercury (Killen et al, 2004). Therefore thermal vaporization is diffusion limited on Mercury.

Photo-ionization with subsequent entrainment of ions in the solar wind is the dominant loss process for alkalis at Mercury. Species such as Ca may be produced hot after dissociation of a sputtered molecule in the atmosphere (Killen, Bida, and Morgan, 2004). Other hot atoms may include Mg. Radiation pressure causing acceleration beyond escape velocity probably contributes to the loss. The rate of photo-ionization is approximately equal to half of the total column in sunlight divided by the ionization lifetime (Killen et al, 2004). Rates of photo-ionization are probably extremely variable, and vary from species to species depending on the differences in gyro-radii and neutrals distributions.

Solar proton radiation induces a process which is termed Photon Stimulated Desorption (PSD). Through this mechanism, individual photons with wavelengths in the UV and EUV ranges are absorbed at the surface and cause the release of particles due to localized electronic excitations. This mechanism is mostly responsible for the release of Na and K from the surface but can also be responsible for the release of other volatile with the planetary surface. When the energy of the impacting particle is on the order of 1 keV, the sputter yield can be of the order of unity. Sputtering is a highly energetic process: the energy characteristic of the substances such as water and sulfur (speculated to exist in permanently shadowed craters at the poles, see Chapter 4 on the Surface). PSD results in the release of particles with characteristic energies that are, somewhat, higher than thermal in a gaussian distribution with a high energy tail (Weibull Distribution), and, correspondingly, somewhat higher scale heights result. Electron Stimulated Desorption (ESD) is a process very similar to PSD in that it too involves particle release due to localized electronic excitations, but fewer energetic electrons are available, so ESD does not play a major role.

Ion or chemical sputtering (the release of atoms and molecules from the surface layer due to momentum transfer or electronic excitation, and/or chemical reaction of implanted solar wind or magnetospheric ions with surface minerals) plays a role at locations where energetic particles interact directly with the surface. Although solar wind ions are in the required energy range and magnetospheric ions can also be accelerated to these energies in the tail, most magnetospheric ions (i.e., photo-ions) do not have sufficient energy to sputter (Killen et al, 2004). Sputtering would result in scale heights being achieved that exceed those due to other release processes except meteoritic vaporization which is the highest temperature energetic process.

Increased ion sputtering can result from ions entering through the cusp regions under favorable solar wind conditions and this is the probable mechanism leading to large and rapid increases in the sodium content of Mercury's exosphere and to the observed high latitude enhancements in Na (Killen et al, 1990; Potter and Morgan, 1990).

Chemical reactions induced by interaction of solar wind hydrogen with oxygen in the regolith would result in the reduction of iron to Fe^0 (metallic iron) on grain boundaries, also known as space weathering. This process is accompanied by the disappearance of iron in the form of Fe^{2+} observable by Near IR spectral reflectance and the resulting difficulty in determining iron abundance (See Chapter 4).

Impact Vaporization is caused by the impact of micro-meteorites and meteorites into Mercury's surface, micrometeorites with a modal speed of 20 km/s (Cintalla, 1992) and meteorites with larger peak velocities of 30 to 40 km/s (Marchi et al, 2004). Because these projectiles can produce a crater having a diameter between one and ten times their own size (depending on the density and porosity of the impactor), material from below the topmost surface is associatively released into space. The vapor probably has a Maxwellian velocity distribution in the range 2500 to 5000 K (Schultz, 1988). In addition, meteoritic infall is a major source of water (Killen and Ip, 1999), though only a very small fraction would end up in polar cold traps. Most trapped water would probably come from large impactors, including asteroids and comets (Killen et al, 1997; Moses et al, 1999).

Direct injection sources would include the solar wind and diffusion from the interior. The solar wind is known to be a source of H, He and noble gases. The interior could be a source of gases, particularly noble gases, such as ^{4}He and ^{40}Ar, or even other volatiles, as suggested by Sprague (1990).

Diffusion is thought to have a role in near surface processes, involving migration of ions following implantation and surface charging, as well as in the transport of volatiles such as water (Sprague, 1992; Madey et al, 1998; Killen et al, 1997; Killen et al, 2004).

The different processes described above not only have characteristic energies, but can be either uniform or variable with local time and location (Lammer et al. 2003; Wurz and Lammer, 2003). Thermal release and PSD correlate with daytime conditions and with the latitude of the release site. On the other hand, sputtering and ESD are only active at those places and times where particles precipitate onto the surface. Impact vaporization is spatially uniform to first order, but depends on the orbital distance which varies from 0.306 to 0.465 AU (Cintala, 1992). For the larger impacts, the individual events are very localized and the associated gas release is transient.

Killen and Ip (199) have pointed out that less energetic processes, PSD and ESD, would remove from the surface the most volatile species only, and would be non-stoichiometric from a surface composition standpoint. On the other hand, more tightly bound species, particularly refractories, would require more energetic processes for removal, such as ion sputtering or micrometeorite impact vaporization, would be stoichiometric from a surface composition standpoint.

5.10 PROPOSED SOURCE AND LOSS PROCESSES

Constituents are initially added to Mercury's exosphere when removed from the surface by a variety of energetic processes, which include hypervelocity micrometeorite impacts, photon-stimulated desorption, and ion sputtering (Killen et al, 2004). The exosphere itself is rapidly eroded, with a lifetime of a few hours, primarily through photoionization (Killen et al, 2004).

The solar wind is the major source of H. The He and Ar exospheres are thought to be derived from a combination radiogenic species from the crust and solar wind implanted species diffused to the surface from the regolith (Hodges, 1974). ^{40}Ar and ^4He are supplied by radiogenic decay and ^{36}Ar and ^3He by solar wind implantation (Killen and Ip, 1999).

The origin of the sodium exosphere was first thought to be the result of photon-stimulated desorption (McGrath et al, 1986), the desorption of neutrals or ions as a direct result of photon electronic excitation of a surface atom.

Using different approaches, Cintala (1992) and Morgan and coworkers (1988) came to the conclusion that vaporization from micrometeorite is likely to be the dominant mechanism for supplying the observed Na and would be entirely sufficient with either efficient recycling of photo-ions back to the surface or greater than lunar abundance of Na or both (Killen and Ip, 1999). Killen et al (2004) have shown that ions are recycled to the surface with an efficiency greater than 50%.

Current thinking is that Na abundances appear to be consistent with a mixture of sources including meteoritic vaporization and photon-induced sources at low latitudes and ion sputter at high and mid-latitudes (Killen and Ip, 1999). Although the emissions for sodium and potassium appear to be generally correlated with one another in distribution, potassium varies at a rate almost twice that of sodium (Potter et al, 2002). The underlying cause of the variable N/K ratio in the atmosphere is currently unknown. The loss rate of Na + K ions is comparable (Killen et al, 2004), but the diffusion rate of K + Na from the interior of surface grains is quite different. Therefore the source ratio may differ, especially following a particularly energetic event that extremely depletes the surface of alkalis. The principal loss mechanism for both ions is thought to be capture by the solar wind. The atoms are not sufficiently different in radiation acceleration losses to explain the high Na/K ratio (Potter et al, 2002).

Na/K ratios appear to be independent of longitude of the subearth (observing) point, demonstrating no correlation with surface rock composition (Potter et al, 2002, 2005; Killen et al, 2004) and potentially indicating that the surface composition doesn't vary or that surface processes which affect exospheric constituent abundances are not influenced by composition. Unlike Mercury's exosphere, which exhibits features at high latitudes, the lunar exosphere, lacking solar wind/magnetosphere interactions, decreases systematically as a function of solar zenith angle.

Enhancements in ion production are likely to be the result of an energetic process, sputtering under special conditions (Killen and Ip, 1999). High latitude enhancements are caused by ion sputtering (Potter and Morgan, 1990). Enhancement right at the subsolar point, called the 'sodium fountain' may be the result of the more energetic chemical sputtering (Potter, 1995).

The dominant sink for all of the ions is now thought to be photo-ionization. Following this process, about half of the ions are recycled to the surface and neutralized (Killen et al, 2004), although a significant fraction is swept up in the flows of the magnetosphere and solar wind. In the case of water, dissociation dominates and thereby provides a substantial source of H, H_2 and O. Solar radiation pressure constitutes a large effect, especially for Na and K, but its role as a sink is probably small, except in the case of unusually fast atoms. Photo-ionization rates estimated vary depending on the observations used and assumptions made about typical conditions (Samson, 1982; Chang and Kelly, 1975; Huebner et al, 1992; Combi et al, 1997). Rates would differ for the active and quiet sun, and vary with orbital distance by up to a factor of 10. Implications are that the steady state sodium supply can be balanced with photo-ionization as the dominant loss mechanism provided that the abundance of sodium in Mercury's regolith is equal to or greater than the abundance of sodium in the Moon's regolith.

The efficiency of photon induced effects, including PSD and thermal release, vary considerably depending on the degree of adsorption of the constituent (Killen and Ip, 1999). In tests done on the thermal release of Na from quartz and corundum films, the release temperature increases considerably during prolonged bombardment (Madey et al, 1998; Shao and Paul, 1993). This finding indicates depletion of more weakly bound (adsorbed as opposed to chemisorbed) species and suggests the ongoing

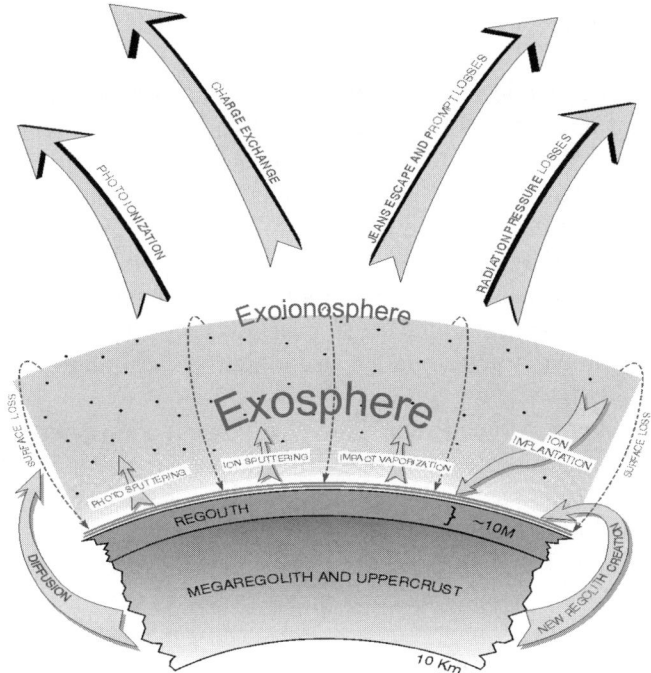

Figure 5-5. Mercury Exosphere Processes, Sources, and Sinks. (Courtesy of Rosemary Killen). **See Color Plate 1.**

depletion of alkalis in equatorial (higher insolation) surfaces and their migration poleward (Leblanc et al, 2004).

Figure 5.5 (courtesy of R. Killen) provides a schematic drawing of the source and sink processes with respect to Mercury's surface (Killen et al, 2004). The total gas pressure at the Hermean surface is only about 10^{-10} mbar (Fjelbo et al. 1976) and the total column density is probably about 10^{12} cm^{-2}. Initial energy distribution of sources of atmospheric components is not influenced by energy distribution of surface components, and vice versa, and thus maintaining detailed balance of the system is not possible (Killen and Ip, 1999).

5.11   MODELS OF MERCURY'S ATMOSPHERE

Providing a dynamic model of Mercury's exosphere is challenging. The distribution of observed species on Mercury is not consistent with a loss-free exosphere accommodated to surface temperature predicted prior to the Mariner 10 encounter. In that case, column abundance would increase as the insolation, temperature, and the hop length, decreased. Instead, the opposite condition is generally observed, where column densities decrease as the nightside is approached. Thus, migration time is considerably longer than ionization time except for the high energy tail of the distribution, and latitudinal variation is great relative to longitudinal variation on short time scales.

Most exospheres, which extend above collisional atmospheres, consist of components with truncated Maxwell-Boltzmann energy distributions, and earlier models made such assumptions in modeling Mercury's atmosphere (Smith et al, 1978; Hodges, 1980). But Mercury has a surface bounded exosphere, and no known process in such an exosphere will result in such a distribution. Component production and survival is the result of complex interactions between source processes and surface interactions, and must be modeled using distributions with peaks which correspond to the binding energy involved in sputtering (Sigmund, 1981; Thompson, 1968; Madey et al, 1998), to the source temperature for PSD, ESD, or impact volatilization (Madey et al, 1998). Other effects, such as dissociation of molecules which result in hot atoms, depend on surface chemistry. These processes are akin to the production of hot atoms in the Martian atmosphere (Fox, 2004).

In addition, Mercury has large orbital eccentricities, which result in large variations in source intensities (photon, charged particle, meteoritic) that in turn result in large variations in radiation pressure, and acceleration of individual constituents (Killen and Ip, 1999).

Smyth and Marconi (1995) used a Monte Carlo approach which combined the best available information concerning the sources, sinks, gas-surface interactions and transport dynamics of sodium and potassium in the exosphere of Mercury. Gravitational attraction as well as radiation pressure acceleration were considered. They considered atomic motions of atoms that initially had velocities >2 km/sec, a small portion of the exospheric constituents (**Figure 5.6**). They assumed a variety of source distributions which vary with orbital distance, ranging from constant to the inverse square of the distance, thus representing a range of processes.

Smyth and Marconi (1995) examined the spatial distribution and relative importance of the initial source atom atmosphere, and the ambient atmosphere. They inferred that variation of the high energy tail of the exosphere distribution was controlled by the extremely large and variable solar radiation acceleration experienced by sodium and potassium as they

SODIUM AND POTASSIUM ATMOSPHERES OF MERCURY

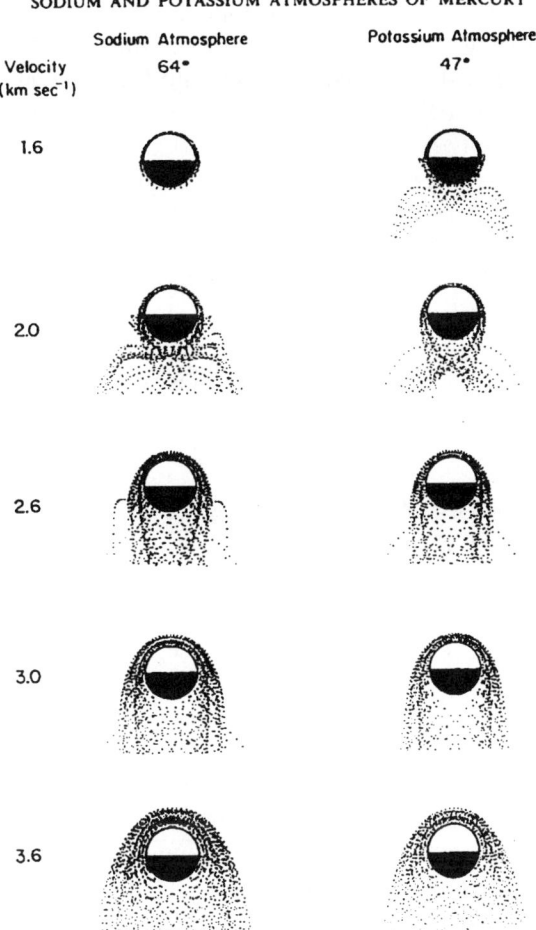

Figure 5-6. Model Spatial Distribution of sodium (left) and potassium (righ). Source populations in Mercury's atmosphere (Smyth and Marconi, 1995) at maximum solar radiation acceleration for a range of ejection speeds. Details of the model are described in the text.

resonantly scatter solar photons. The lateral, anti-sunward, transport rate of high energy sodium and potassium ambient atoms was shown to be driven by the solar radiation acceleration, and to form a tail. Over a significant portion of Mercury's orbit about the Sun, solar radiation acceleration is rapid enough to compete with the relatively quick photo-ionization process for these atoms when they are located on the sunlit surface near, or within 30^0 of the terminator. The lateral transport rate derived from model calculations depends on a migration time for an ensemble of atoms, initially starting at a point source on the surface. The best model fits to the observational data require a significant degree of thermal accommodation of the ambient sodium atoms to the surface and a source rate that decreases as an inverse power of 1.5 to 2.0 in heliocentric distance.

The model of Smyth and Marconi (1995) did demonstrate the kind of variability that agreed in a general way with the nature of observed variability. However, one to one correspondence between model and observations can't be demonstrated until observations can be taken over a complete solar cycle. In addition, great local and short-term variations are observed, and generate ambiguity about what is 'typical'. Mercury's environment is so dynamic that, to date, no predicted dynamic model has consistently fit the observations for any species (Killen and Ip, 1999).

5.12   SUMMARY OF CONSTITUENT SOURCE AND LOSS MECHANISMS

Generally, the sources of the exospheric constituents at Mercury presumably include 1) the regolith, 2) the interior (He, Ar, Rn), 3) evaporation of incoming meteorites (and an occasional comet), and 4) implantation of solar wind ions. The overall atmospheric dynamics are complicated by the very effective cold trap provided by Mercury's night-side (Curtis and Hartle, 1978) and polar craters (Harmon et al, 1990). Atmospheric constituents are eventually lost when they become ionized by either solar UV or charge exchange, although they may undergo many recycling events before being thus lost (Killen et al, 2004).

The role of the solar wind as a source for Mercury's light atom atmosphere, particularly for H and He, has been predicted and observed and is relatively well understood (Hunten et al, 1988). However, because we have only two snapshots of the H, He and O exospheres from Mariner 10, the true spatial distribution of these gases and their relationships to the state of the magnetosphere and solar wind are still essentially unknown.

The exospheres of each neutral species (including Na and K) observed later from ground-based telescopes are known to extend well beyond the front-side magnetopause in a way that is unique to each constituent. However, knowledge of the principal source and loss processes involved is incomplete even for sodium, the most abundant and most observed constituent.

Ca, Mg and Al are signature species for sputtering because their binding energies are too large to be promoted effectively to the exosphere by less energetic processes (photon-stimulated desorption and thermal evaporation). Thus, the distributions of these species should reveal much more about exospheric processes than can be learned by studying only the distribution of

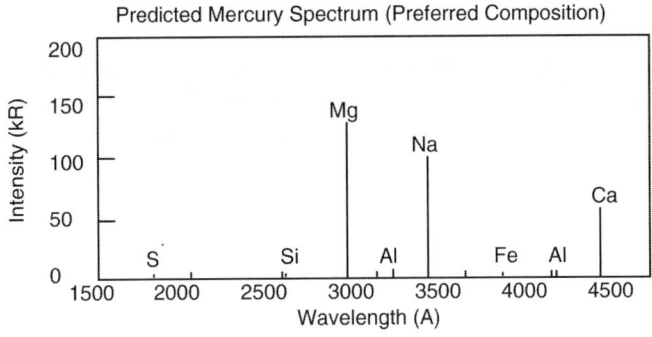

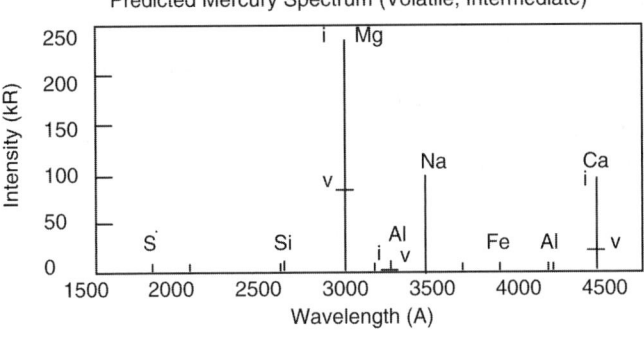

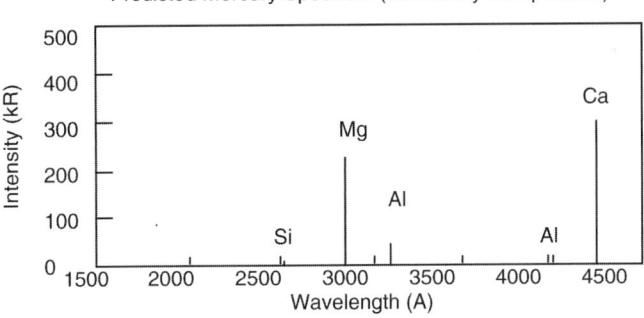

Figure 5-7. Predicted Atmosphere as Function of Mercury Bulk Abundances for, from top, a) best predicted from observations (Morgan and Killen, 1997), b) volatile (V), intermediate (I), and c) refractory compositions. Reproduced from Morgan and Killen (1997) with permission of Elsevier Science.

sodium (remembering that this element can be elevated above the surface by a number of different processes).

Mercury exospheric abundances of refractories, such as calcium, are potential indicators of planetary bulk composition as illustrated in **Figure 5.7** (Killen and Ip, 1999). Refractories in the atmosphere could vary by factors of two to four between volatile and intermediate source compositions, and an order of magnitude between intermediate and refractory source compositions (Goettel, 1988). However, abundance measurements are highly temporally variable on the same scale on Mercury. A time series of observations would need to be taken to separate temporal variations from compositional influences.

5.13   MERCURY'S EXO-IONOSPHERE

Photoelectrons, solar wind ions, sputtered ions and electrons, and even plasma accelerated toward Mercury in its tail may constitute a conducting zone known as an ionosphere. The density and dynamics of such a region are not well known, but Mariner 10 did reveal that Mercury does not have an Earthlike ionosphere with permanent radiation belts and a plasmasphere (Killen and Ip, 1999). Nonthermal loss by acceleration of ionized components in the ionosphere could be a major influence on the rate of nonthermal erosion of the atmosphere, which depends on the as yet poorly known density and structure of the ionosphere.

Some have argued that Mercury has essentially no conducting ionosphere (Christon, 1989), others that the regolith, not an ionosphere, carries conducting current (Southwood, 1996). The surface could become electrostatically charged through mechanisms such as sputtering and result in large current generation (Grard, 1997).

The extent of surface charging or ionization depends on the extent to which the magnetosphere shields the surface. The magnetosphere standoff distance is known to vary from surface, providing no shielding, to about 1 to 1.5 planetary radii. Thus the behavior of the magnetosphere and its interaction with the solar wind has major influence on the creation and loss of Mercury's atmosphere.

Magnetosphere/ionosphere/solar wind interactions are discussed in detail in Chapter 6.

5.14   SPACE WEATHERING AS AN ATMOSPHERE
        MODIFICATION PROCESS

Space weathering processes include the effect of magnetospheric particles on atmospheric neutrals, as well as the effect of photo-ionized atmospheric species on surface composition. Such interactions are known to result in observed spatial and temporal variations in the distribution of atmospheric species (Killen et al, 2004; Leblanc et al, 2004). Studies of space weathering involve experimental, observational, and theoretical work, and have been a major thrust of those who study Mercury's exosphere (Killen et al, 2004) and surface (See Chapter 4 on the Surface). A special emphasis has been on understanding the role of the magnetosphere (See Chapter 6 on the Magnetosphere) in space weathering.

Because Mercury has no atmosphere to protect its surface, various kinds of external radiation interact directly with its surface (including solar photon radiation; energetic ions from Mercury's own magnetosphere; the solar wind; galactic cosmic rays and micro-meteorites). Solar wind plasma has limited but variable access to the surface because Mercury's magnetosphere is able to provide certain shielding (Massetti et al, 2003; Kallio and Janhunen, 2003; Sarantos et al, 2001; Kabin et al, 2000).

The three basic types of magnetospheric models (to be described in much greater detail in Chapter 6 on the Magnetosphere) include an analytical model (Sarantos et al, 2001), a quasi-neutral hybrid (Kallio and Janhunen, 2003), and MHD models (Kabin et al, 2000; Ip and Kopp, 2002; Massetti et al, 2003). Killen and coworkers (2004) performed simulations and modifications of available magnetospheric models (Toffoletto and Hill, 1993; Ding et al, 1996; Sarantos et al, 2001) for conditions anticipated for Mercury. Generally, they were able to reproduce Mariner 10 measurements. The complex hybrid model (Kallio and Janhunen, 2003) seems to be in qualitative agreement with observations.

All models have indicated that Mercury's magnetosphere could be open to the solar wind over substantial areas when the interplanetary magnetic field turns in a southward direction (e.g. Slavin, 2004; Kabin et al, 2000; Sarantos et al, 2001). A strong magnetic field component (Bx) would introduce a clear asymmetry in the north-south and east/west directions (Kallio et al, 2004; Sarantos et al, 2001). Preferential precipitation would be expected in the north when Bx is strongly negative and in the south when Bx is strongly positive (Sarantos et al, 2001). In these circumstances, solar wind protons and accelerated heavier ions could reach Mercury's regolith and act as sputtering agents for exospheric species, as indicated in **Figure 5.8** (Slavin et al, 2004). These release processes would affect only the topmost atomic layers of the surface, in contrast to micrometeorite impact which would release material from greater depths (Cremonese et al, 2004). Sputtering events are twice as likely to occur in the southern hemisphere (Potter et al, 2006) for reasons that can't be explained based on our current knowledge of solar wind/magnetosphere interaction.

The models differed in predicting the extent and intensity of interactions with the surface (Killen et al, 2004). Spatial variations in the atmosphere are known to be affected by solar wind induced ion sputtering. Although Mercury's magnetic field normally deflects the solar wind, direct penetration through the cusps can result in ion sputtering of neutral atoms (Ip, 1986). Such an event occurs either when field lines are open through the cusps (Luhmann et al, 1998; Sarantos et al, 2001) or when unusually high solar

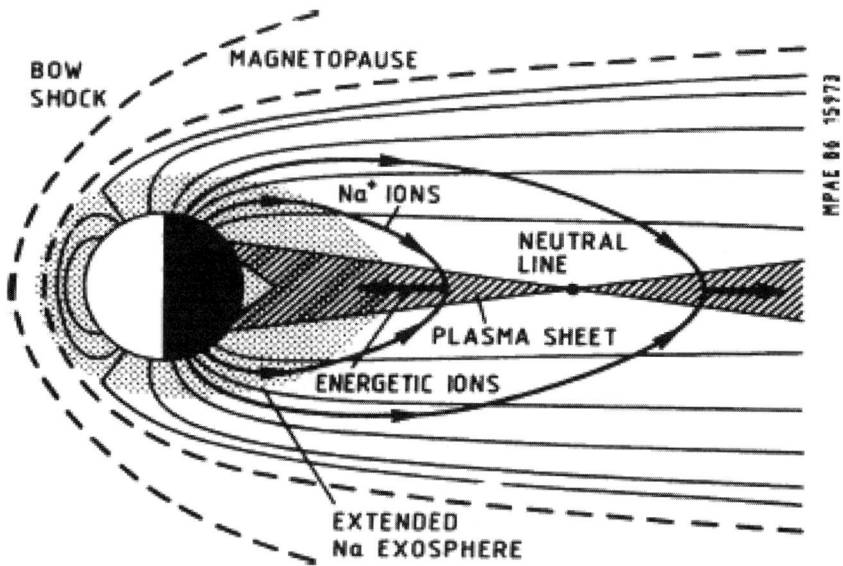

Figure 5-8. Role of the Magnetosphere in Sodium Distribution. (Reprinted from Slavin, 2004, with permission of Elsevier.) Recirculation and return of sodium sputtered off of the surface via magnetospheric convection.

wind pressure pushes the magnetopause to the surface (Goldstein et al, 1981; Kabin et al, 2000). Following storms and substorms, wide auroral bands accompanied by ion sputtering could extend into the dayside.

Recycling of ions is needed to account for observed steady state abundances of atmospheric species (Killen and Ip, 1999). In order to account for ion recycling, Killen and coworkers (2004) developed a full-particle tracing code allowing the trajectories of 3500 Na ions to be followed under combined gravitational, magnetic, and electrical forces as they were generated and impacted the surface (seconds), crossed the magnetopause and escaped (tens of seconds), or moved down the tail and reimpacted the nightside (minutes).

Killen and coworkers (2004) demonstrated that 60% of the ions neutralize and are retained on surface impact, but with a characteristic east/west asymmetry in the retention pattern which results from the assumed dawn-dusk electric field. Ions launched on the dawnside are much more likely to be retained, those launched on the duskside much more likely to be lost the solar wind. A weaker north/south asymmetry can be observed due to strong negative Bx of the IMF, increasing the likelihood that volatiles are deposited at higher latitudes in polar cold traps.

Shown in **Figure 5.9** are the resulting distributions of ion recycling ratios predicted from current modeling of space weathering. Recycling is indeed demonstrated to be important: 60% of launched ions are neutralized and

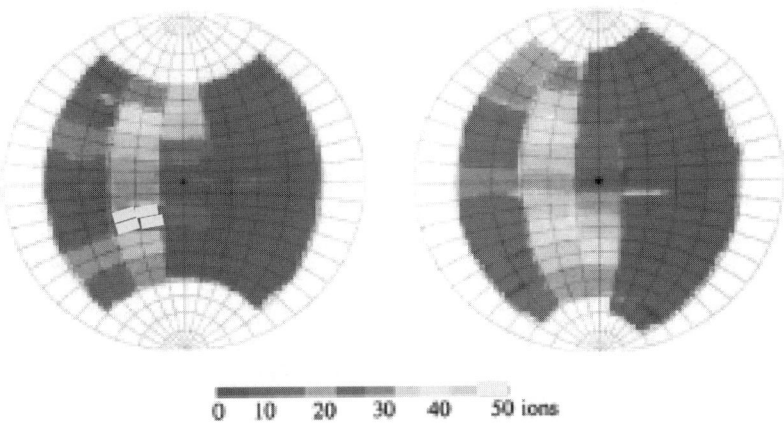

Figure 5-9. Predicted Global Ion Recycling showing ions returning to dayside surface. Ions are generated according to double cosine law. Ion turnover rate and area segmentation depends on the choice of IMF. Dawn is on the left, dusk on the right. (Reprinted from Killen et al, 2004, with permission of Elsevier.)

retained upon impacting the surface. A fraction of redeposited ions have sufficient energy to produce neutral sputtering instead of neutralizing on impact. This result is in agreement with earlier estimates and earlier assessments that abundances are regulated by the IMF (Killen et al, 2001). The IMF is thought to control atmospheric abundances via magnetic forces through retention of ions that would have been lost otherwise and by augmenting neutrals contribution along open field lines. Ions escaping to the solar wind as well as those neutralized by reimpact of the dayside surface (and thus no longer reusable) are considered losses. Features on the Figure 5.9 images include dawnside enhancement, which matches observations. The characteristic east/west asymmetry in the retention pattern results from the assumed dawn-dusk electric field. Ions launched on the dayside are much more likely to be retained, while those launched on the duskside much more likely to be lost the solar wind. A weaker north/south asymmetry can be observed due to strong negative Bx of the IMF, increasing the likelihood that volatiles are deposited at higher latitudes in polar cold traps. (Sarantos et al, 2001; Kallio and Janhunen, 2003).

It should be noted that the difference in Na distribution patterns on the Moon and Mercury could be explained by the lack of magnetosphere and weaker solar wind interactions on the Moon. Whereas the Na atmosphere is highly irregular on Mercury, on the Moon Na distribution is relatively uniform, with no high latitude enhancements. This could be explained by the lack of magnetospheric effects on the Moon as well as the weaker solar flux and solar wind. On the other hand, the uniform distribution of Na in lunar surface rocks could also be a contributing factor.

5.15 SUMMARY

The only measured constituents of Mercury's atmosphere are H, He, O, Na, and K, all of which were detected optically from resonantly scattered sunlight (Broadfoot et al, 1976). Species such as Ca, Al, Fe, Mg, Si, S, and OH are predicted constituents (Morgan and Killen,1997). Mercury's atmosphere is an observable result of a dynamically coupled system, with the planetary surface acting as both source and sink, and the magnetosphere as both buffer and accelerator for solar wind plasma. The low density of volatiles from the surface and the small internal magnetic field combine to produce this unusual environment, where the planetary surface replenishes a true exobase. Based on the observed thermal to gravitation potential energy ratio, Mercury should be surrounded by a quasi-spherical region, 100,000 km (~40 RM) in diameter, of ionized atmospheric constituents that leave the planet as neutrals and are subsequently ionized by the solar wind and solar UV photons. These ions are then picked up and convected downstream by solar wind V x B forces (Wu et al, 1973). A more complete understanding of this unusual planetary exosphere has great significance for comparison with a broad range of objects in the solar system. The flux and energy of the ions reaching the orbit of Mercury provide a variable source of plasma. Both the content and morphology of the exosphere are modulated by changes in the solar wind, which itself interacts with the magnetosphere.

5.14 REFERENCES

Antoniadi, Eugene, *La planete Mercure et la Rotation des Satellites*, (Publ. Paris, Gauthier –Villars) 1934.

Banks, P.M., H.E. Johnson, and W.I. Axford, The atmosphere of Mercury, *Astrophys. Space Phys.*, **2**, 214–220, 1970.

Bida, T.A., R.M. Killen, and T.H. Morgan, Discovery of calcium in Mercury's atmosphere, *Nature* **404**, 159-161, 2000.

Broadfoot, A.I., S. Kumar, M.J. Belton, and J.B. McElroy, Mercury's atmosphere from Mariner–10. Preliminary results. *Science* **185**, 166–169, 1974.

Broadfoot, A.I., J.E. Shemansky, and S. Kumar, Mariner-10 Mercury atmosphere, *GRL,* **3**, 577–580, 1976.

Chamberlain, J.W. and D.M. Hunten, *Theory of Planetary Atmospheres*, 2nd Ed. (Publ. Academic Press), 481 p., 1987.

Chang, J., and H. Kelly, Photoabsorption of the neutral sodium atom: a many-body calculation, *Phys Rev A*, **12**, 92-98, 1975.

Cheng, A., R. Johnson, S. Krimigis, L. Lanzaroti, Magnosphere, exosphere and surface of Mercury, *Icarus,* **71**, 430-440, 1987.

Christon, S., 1987, A comparison of the Mercury and Earth magnetospheres: electron measurements and substorm time scales, *Icarus*, **71**, 448-471, 1987.

Cintala, M.J., Impact induced thermal effects in the lunar and Mercurian regoliths, *JGR*, **97**, 947–973, 1992.

Combi, M., M. DiSanti, and U. Fink, The spatial distribution of gaseous atomic sodium in the comae of comets: evidence for direct nucleus and extended plasma sources, *Icarus*, **130**, 336-354, 1997.

Cremonese G., M. Capria, V. Achilli, F. Angrilli, P. Baggio, C. Barbieri, J. Baumgardner, N. Bistacchi, F. Capaccioni, A. Caporali, I. Casanova, S. DeBei, G.Farlani, S. Fornaier, D. Hunten, W. Ip, M Lazzarin, I. Longhi, L. Marinangeli, F. Marzari, P. Massironi, P. Masson, M. Mendillo, B. Pain, G. Preti, R. Ragazzoni, J. Taitala, G. Salemi, M. Sgavetti, A. Sprague, E. Suetta, M. Tordi, S. Verani, J. Wilson, L. Wilson, MEMORIS: a wide angle camera for the BepiColombo mission, *Advances in Space Research*, **33**, Mercury, Mars and Saturn, 2182-2188, 2004.

Curtis, S.A. and R. Hartle, Mercury's Helium exosphere after Mariner 10's third encounter, *JGR Space Physics*, **83**, 1551-1557, 1987.

Ding, C., T. Hill, B. Ramaswamy, Modeling and mapping of electric potential on closed field lines, in *Physics of Space Plasmas*, #14, Ed., T. Chang, J. Jasperse, MIT center for theoretical Geo/Cosmo Plasma Physics, Cambridge, MA, 645, 1996.

Fink, U., H.P., Larson, and R.F. Popper A new upper limit for an atmosphere of CO_2, CO on Mercury *Astrophys J*, **187**, 407-415, 1974.

Fjelbo, G.A. and F. Kilore The occultation of Mariner 10 by Mercury *Icarus*, **29**, 407-415, 1976.

Flynn, B. and M. Mendillo, Simulations of the lunar sodium atmosphere, *JGR*, **100**, 23271-23278, 1995.

Goettel, K., Present bounds on the bulk composition of Mercury: Implications for planetary formation processes, in *Mercury,* Ed. Vilas, Chapman, Metthews, 613-621, U. Arizona Press, Tucson, 1988.

Goldstein, B., S. Suess, R. Walker, Mercury: Magnetospheric processes and the atmospheric supply and loss rates, *JGR*, **86**, 5845-5899, 1981.

Grard, R., Photoemission on the surface of Mercury and related electrical phenomena, *Plan Space Sci*, **45**, 67-72, 1997.

Hartle, R., K. Ogilvie, C. Wu, Neutral and ion-exospheres in the solar wind with applications to Mercury, *Plan Space Sci*, **21**, 2181-2191, 1973.

Hartle, R.E., S. A. Curtis, G.E. Thomas, Mercury's helium exosphere, *JGR*, **80**, 3689-3692, 1975.

Hartle, R.E. and R.M. Killen, Measuring pickup ions to characterize the surfaces and exospheres of planetary bodies: Applications to the Moon, *GRL*, **33**, 5, L05201, 2006.

Hodges, R., Model atmospheres for Mercury based on a lunar analogy, *JGR.*, **79**, 2881–2885, 1974.

Hodges, R., Methods for Monte Carlo simulation of the exospheres of the Moon and Mercury, *JGR*, **85**, 164-170, 1980.

Huebner, W.F., J.J. Keady and J.P. Lyon, Solar photo rates for planetary atmospheres and atmospheric pollutants, *Astrophys. Space Sci.* **195**, 1–294, 1992.

Hunten, D.M., T.H. Morgan and D.H. Shemansky, The Mercury Atmosphere in *Mercury,* Eds. Vilas, Chapman, and Matthews, (Publ. Univ. of Arizona, Press, Tucson, 562–612, 1988.

Ip, W., The sodium exosphere and magnetosphere of Mercury; *GRL*, **13**, 423–426, 1986.

Ip, W., Dynamics or electrons and ions in Mercury's magnetosphere, *Icarus,* **71**, 441-447, 1987.

Kabin, K. and T. Gombosi et al, Interaction of Mercury with the solar wind, *Icarus,* **143**, 397-406, 2000.

Kallio, E. and P. Janhunen, The response of the Hermean magnetosphere to the interplanetary magnetic field, *Adv in Space Research,* **33**, 2176-2181, 2004.

Killen, R.M., J. Benkhoff, and T.H. Morgan, Mercury's polar caps and the generation of an OH exosphere, *Icarus,* **125**, 195–211, 1997.

Killen, R.M., T.A. Bida, T.H. Morgan, Calcium Exosphere of Mercury, *Icarus,* **173** (2), 300-311, 2005.

Killen, R.. and W. Ip, The surface bounded atmospheres of Mercury and the Moon, *Reviews of Geophys.* **37** (3) 361–406, 1999.

Killen, R. and T. Morgan, Maintaining the Na Atmosphere of Mercury, *Icarus,* **101** (2): 293-312, 1993.

Killen, R. and T. Morgan, Diffusion of Na and K in the uppermost regolith of Mercury, *JGR*, **98**, 23589-23601, 1993.

Killen, R.M., A.E. Potter, and T.H. Morgan, Spatial distribution of sodium vapor in the atmosphere of Mercury, *Icarus,* **85**,145–167, 1990.

Killen, R.M., A.E. Potter, A. Fitzsimmons, and T.H. Morgan, Sodium D2 line profiles. Clues to the temperature structure of Mercury's exosphere, *Planet. Space Sci.* **47**, 1449–1458, 1999.

Killen RM, A.E. Potter, P. Reiff, M. Sarantos, B.V. Jackson, P. Hick, B. Giles, Evidence for space weather at Mercury, *JGR Planets* **106** (E9): 20509-20525, 2001.

Killen, R.A., M. Sarantos, P. Reiff, Space Weather at Mercury, *Adv Space Research*, **33**, 1899-1904, 2004.

Kumar, S., Mercury's atmosphere. A perspective after Mariner 10. *Icarus* **28**, 579–592, 1976.

Laakso, H., H. Koskinen, T. Pulkkinen, R. Grard, Electric current systems in the Ml. Mercury magnetosphere, Abstract, *EOS Trans.* AGU, **78** (46) , Fall Meeting, Fall. Meeting Suppl, 1997.

Lammer, H., P. Wurz, M. Patel, R.M. Killen, C. Kolb, S. Massetti, S. Orsini, A. Milillo, The variability of Mercury's exosphere by particle and radiation induced surface release processes, *Icarus,* **166,** 238-247, 2003.

LeBlanc, F., D. Delcourt, R.E. Johnson, Mercury's sodium exosphere: Magnetospheric ion recycling, *JGR Planets*, **108,** E12, #5136, 2003.

Lewis, J.S., Metal/silicate fractionation in the solar system *Earth Plan Sci Lett,* **15,** 286-290, 1972.

Lewis, J.S., *Physics and Chemistry of the Solar System*, 2nd Ed. (Publ. Elsevier), 655p., 2004.

Madey, T.E., B.V. Yakshinskiy, V.N. Ageev, and R.E. Johnson, Desorption of alkali atoms and ions from oxide surfaces: relevance to origins of Na and K in the atmospheres of Mercury and the Moon, *JGR*, **103,** 5873–5887, 1998.

Massetti, S., S. Orsini, A. Milillo, S. Orsini, A. Mura, E. De Angelis, H. Lammer, P. Wurz,, Mapping of the cusp plasma precipitation on the surface of Mercury, *Icarus,* **166,** 229-237, 2003.

McGrath, M.A., R.E Johnson, and L.J. Lanzerotti, Sputttering of sodium on the planet Mercury, *Nature,* **323,** 694–696, 1986.

Morgan, T.H. and R.M. Killen, A non-stoichiometric model of the composition of the atmospheres of Mercury and the Moon, *Planet. Space Sci.* **45** (1), 81–84, 1997.

Morgan, T.H. and A.E. Potter, Distributions of sodium and potassium vapor on Mercury, *Bull. Amer. Astron. Soc.* 24957, 1992.

Morgan, T.H. and D.E. Shemansky, Limits to the lunar atmosphere, *JGR Planets*, **96,** 1351– 1367, 1991.

Morgan, T.H ., H.A. Zook, and A.E. Potter, Impact–driven supply of sodium and potassium to the atmosphere of Mercury, *Icarus* **75,** 156–170, 1988.

Moses, J., K. Rawlins, K. Zahnle, External sources of water for Mercury's putative ice deposits, *Icarus,* **137**(2), 197-221, 1999.

Ogilvie, K., J. Scudder, V. Vasyliunas, R. Hartle, G. Siscoe, Observations at the planet Mercury by the plasma electron experiment: Mariner 10, *JGR,* **82,** 1807-1824, 1977.

Pierce, K., Construction of a Bowen image slicer, *Publ. Astron. Soc. Pac.* **77,** 216–217, 1965.

Potter, A.E., in Proc. Workshop on Sodium Atmospheres, Exospheres and Coronae in the Solar System (*Publ. Sun Juan, California*), **25,** 1993.

Potter, A.E., Chemical sputtering could produce sodium vapor and ice on Mercury, *GRL,* **22,** 3289-3292, 1995.

Potter, AE, C. Anderson, R. Killen, T. Morgan, Ratio of sodium to potassium in the Mercury exosphere, *JGR Planets*, **107**, doi:10.1029/2000JE001493, 2002.

Potter, A. E. and T.H. Morgan, Discovery of sodium in the atmosphere of Mercury, *Science*, **229**, 651–653, 1985.

Potter, A.E. and T.H. Morgan, Potassium in the atmosphere of Mercury, *Icarus*, **67**, 336–340, 1986.

Potter, A.E. and T.H. Morgan, Variation of sodium on Mercury, with solar radiation pressure, *Icarus,* **71**, 472–477, 1987.

Potter, A.E. and T.H. Morgan, Discovery of sodium and potassium vapor in the atmosphere of the Moon, *Science*, **241**, 675-680, 1988.

Potter, A.E. and T.H. Morgan, Evidence for magnetospheric effects on the sodium atmosphere of Mercury, *Science*, **248**, 835–838, 1990.

Potter, A.E. and T.H. Morgan, Evidence for suprathermal sodium atmosphere of Mercury, *Adv. Space Res.*, **19** (10), 1571–1576, 1997.

Potter, A. and T.H. Morgan, Sodium and potassium atmospheres of Mercury, *Planet Space Sci*, **45**, 95-100, 1997.

Potter, A.E., R.M. Killen, and T.H. Morgan, Rapid changes in the sodium exosphere of Mercury, *Planet. Space Sci.,* **47**, 1441–1448, 1999.

Potter, A.E., R.M. Killen, and T.H. Morgan, The sodium tail of Mercury, *Meteor.and Planet. Sci,.* **37**, 1165– 1172, 2002.

Potter, A.E., R.M. Killen, and M. Sarantos, Spatial distribution of Sodium on Mercury, *Icarus*, **181** (1), 1-12, 2006.

Samson, J., Atomic photoionization, in *Encyclopedia of Physics*, **31**, Ed S. Flugg, 123 -213, Springer-Verlag, New York, 1982.

Sarantos, M., P. Reiff, T. Hill, R. Killen, A. Urquhart, A Bx interconnected magnetosphere model for Mercury, *Planet Space Sci*, **49**, 1629-1635, 2001.

Shao, Y. and J. Paul, TPD studies of the interaction of D2O and Na with clean and oxidized Al (100) surfaces, *Appl Surf Sci*, **72**, 113-124, 1993.

Schultz, P., Cratering on Mercury: A relook. In *Mercury*, Vilas, Chapman, Matthews, Eds., U. Arizona Press, 274-335, 1988.

Shemansky, D.E. and T.H. Morgan, Source processes for the alkali metals in the atmosphere of Mercury, *GRL*, **18**, 1659–1662, 1991.

Shemansky, D., Revised atmospheric species abundances at Mercury: The debacle of bad g values, *The Mercury Messenger*, **1**, Issue 2, 1988.

Shemansky, D.H. and A.I. Broadfoot, Interaction of thee surfaces of the Moon and Mercury with their exospheric atmospheres, *Rev. Geophys*, **15**, 491–400, 1977.

Sigmund, P., Sputtering by ion bombardment, in *Theoretical concepts, in Sputtering by particle bombardment* **I**, ed. R. Behrisch, 9-72, Springer-Verlag, New York, 1981.

Slavin, J.A. and R.E. Holzer The effect of erosion on the solar wind stand off distance at Mercury, *JGR Space Physics*, **84**, 2976–2082, 1979.

Slavin, J., Mercury's Magnetosphere, *Adv Space Research*, **33**, 1859-1874, 2004.

Smith, G.R., D.E. Shemansky, A. Lyle, L. Wallace, Monte Carlo modeling of exospheric bodies: Mercury, *JGR Space Physics*, **83**, 3783-3790, 1978.

Smyth, W., Nature and variability of Mercury's sodium atmosphere, *Nature*, **323**, 696-699. 1986.

Smyth, William H. and M.L. Marconi, Theoretical overview and modeling of the sodium and potassium atmospheres of Mercury, *Astrophys. J.,* **441** (2), Part 1, 839–864, 1995.

Sprague, A.L., R.W. Kozlowski, and D.M. Hunten, Caloris Basin: An enhanced source for potassium in Mercury/s atmosphere, *Science*, **249**, 1140–1143, 1990.

Sprague, A.I. , R.W. Koslovski, D.M. Hunten and F.A. Grosse, An upper limit on neutral calcium in Mercury's atmosphere, *Icarus*, **104**, 33–37, 1993.

Sprague, A.I., R.W. Kozlowski, and D.M. Hunten, An upper limit on neutral calcium in Mercury's atmosphere, *Icarus*, **104**, 33–37, 1995a.

Sprague, A.I., D.M. Hunten and R. Lodders, Sulfur at Mercury,   Elemental at the poles and sulfides in the regolith, *Icarus*, **118**, 211–215, 1995b.

Sprague, A.I., D.M. Hunten, and R. Lodders, Sulfur and Mercury: Elemental at the poles and sulfides in the regolith. Erratum, *Icarus*, **123**, p. 247, 1996.

Sprague, A.I., D. Hunten, R. Kozlowski, F. Grosse, R. Hill, R. Morris, Observations of the sodium in the lunar atmosphere during International Lunar Atmosphere Week, 1995, *Icarus*, **131**, 372-381, 1998.

Sprague et al, upper limit for lithium in Mercury's atmosphere, *Icarus,* **123**, 345-349, 1996.

Stern, S.A., A. Fitzsimmons, R.M. Killen, and A.E. Potter, A direct measurement of sodium temperature in the lunar exosphere, in *Lun. Plan Sci. XXXI*, 1122.pdf, 2000.

Thompson, M. Energy spectrum of ejected atoms during high energy sputtering of gold, *Phil. Mag.*, **18**, 306-314, 1968.

Toffoletto, F.R., T.W. Hill, A nonsingular model of the open magnetosphere. *JGR,* **98**, 1339–1344, 1993.

Wurz P, H. Lammer, Monte Carlo Simulation of Mercury's Exosphere, *Icarus*, **164** (1), 1-13, 2003.

Zollner, Johann Karl Friedrich, Photometrische Untersuchungen uber die physische Beschaffenheit des Planetem Mercur (Photometric researches on the physical condition of the planet Mercury) in *Poggendorf's Jubelband* , 1874.

5.15   SOME QUESTIONS FOR DISCUSSION

1. What benefit came from the apparent failure of the UVS experiment on Mariner 10?

2. Why are Na and K observed in Mercury's exosphere?

3. Compare exospheres and atmospheres and where Mercury fits on that spectrum.

4. How would you design an orbital UVS experiment to measure the anticipated variation in profiles of H, Na, and K over the course of one Mercury year?

Chapter 6

MERCURY'S MAGNETOSPHERE

6.1   PRE-MARINER 10 KNOWLEDGE OF MERCURY'S MAGNETOSPHERE

Prior to the Mariner 10 encounters, Mercury was assumed to lack a global magnetic field and, as a result, to have a lunarlike interaction with the solar wind and no significant magnetosphere (Strom and Sprague, 2003). Mercury's location deep in the heliosphere results in solar wind pressure and interplanetary magnetic field intensity during it's perihelion that are an order of magnitude higher than the Earth's. Could these conditions result in the magnetosphere being pushed into the magnetotail and creating aurora-like storms (Kallenrode, 2004)? Mariner 10 was targeted to cross the 'wake' of the planet in order to survey its interaction with the solar wind (Strom and Sprague, 2003). The discovery of a magnetosphere was an unanticipated and serendipitous discovery.

6.2   MARINER 10 MAGNETOSPHERE DETECTION

During Mariner 10's encounters, many structures and behaviors reminiscent of Earth's magnetosphere were observed. The magnetic field is aligned closely with Mercury's rotation axis. Its intensity is approximately 10^{-4} that of the Earth's, but still strong enough to produce a bow shock between the planet and the incoming solar wind and to contain accelerated electrons and protons. There is a magnetopause, magnetotail, plasma sheet (with the familiar electron energy spectra), magnetic field depolarization during multiple substorm-like plasma injections, and quasi-periodic alternating plasma sheet with boundary layer plasma variations near the

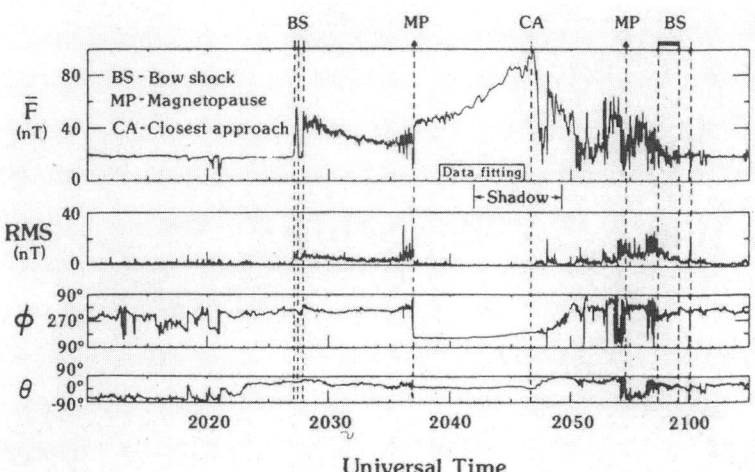

Universal Time

Figure 6-1. Magnetic Field data during the first encounter of Mariner 10 with Mercury on March 29, 1974. The top panel shows the magnetic field magnitude, the next panels show, respectively, the standard deviation of the magnetic field over all three components, the ecliptic longitude (φ) and the ecliptic latitude (θ) of the field. The times of the Bow Shock (BS), Magnetopause (MP) crossings, and Closest Approach (CA) are marked by vertical lines. US indicates the location of upstream waves ((Russell et al, 1988, The Magnetosphere of Mercury, in Mercury, Copyright 1988, The Arizona Board of Regents. Reprinted by permission of the University of Arizona Press.)

magnetopause. The magnetopause ranges from 1.5 to 2.4 planetocentric Mercury radii (R_M) from the planet toward the sun. The polar diameter of the magnetosphere is not directly known. Much of the solar wind appears to be 'stopped' by the magnetic field, except possibly for regional impacts over the planet during solar storms. The magnetopause, generally well off the surface, may descend down to the surface during periods of intense solar activity, particularly near perihelion (Ness et al, 1974, 1975, 1976). Modest radiation belts are suspected.

Mariner 10 flew by Mercury three times in the course of the mission. On the first and third encounters, Mariner 10 approached the planet along northern hemisphere trajectories with low enough periapses to detect magnetospheric boundaries.

Signatures of magnetosphere boundary crossings, first a bow shock and then a magnetopause, were unexpectedly detected in the low energy electron (between 13.4 and 688 eV) measurements some nineteen minutes before reaching Closest Approach and 700 km above the planet's unilluminated hemisphere during Encounter I (Ogilvie et al, 1974). Meanwhile, the charged particle telescope, designed to monitor higher energy electrons (Ee >170 keV) and protons (Ep >500 keV), indicated that bow shock and magnetopause boundaries had been crossed. Similar transitions were in each case identified in the outbound data (Simpson et al, 1974).

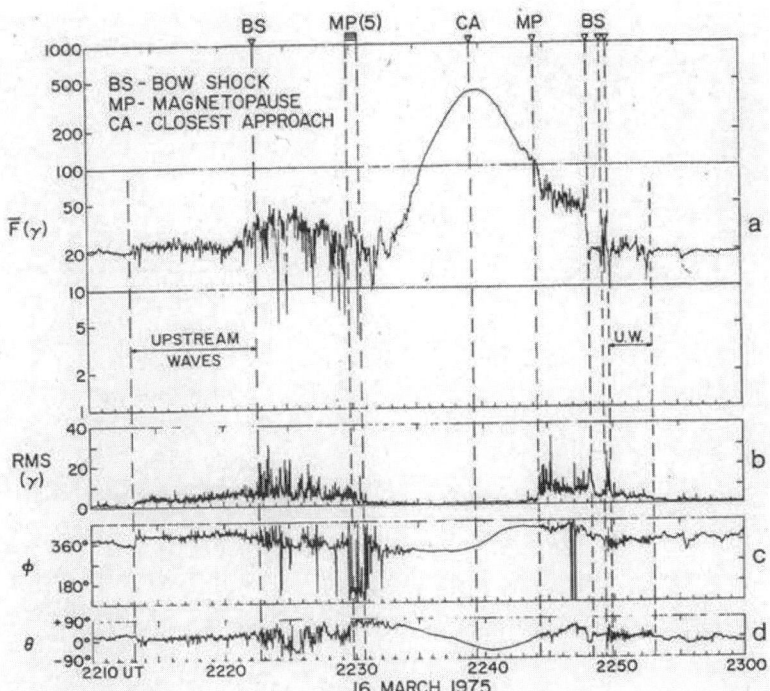

Figure 6-2. Magnetic Field Measurements during Third Encounter on March 16, 1975 (Russell et al, 1988, The Magnetosphere of Mercury, in Mercury, Copyright 1988, The Arizona Board of Regents. Reprinted by permission of the University of Arizona Press). Panels from top to bottom, respectively, show (a) magnetic field magnitude, (b) standard deviation of the field over all three components, (c) ecliptic longitude (φ) and (d) ecliptic latitude (θ).

Magnetic field measurements recorded during Encounter I are presented in **Figure 6-1** (Russell et al, 1988 based on work of Ness et al, 1974). The location inbound of a (quasi-perpendicular) bow shock (BS) is clearly indicated by a sharp localized, rise in field magnitude. Multiple crossings of this boundary were made because it 'flapped' back and forth at velocities greater than that of the spacecraft. The magnetic field signature of the outbound (quasi-parallel) shock was more diffuse. Recognition by the experimenters of these signatures was based on their existing experience of identifying bow shock and magnetopause crossings in the Earth's magnetosphere.

Magnetic field measurements made during Encounter III on 16 March, 1975, can be seen in **Figure 6-2** (Russell et al, 1988 based on work of Ness et al, (1976). The different trajectory of encounter 3, closer and more pole-ward than the Encounter I trajectory (as seen in Figure 6-1) was used to provide confirmation and unequivocal evidence for the global nature of the magnetic field unexpectedly detected during the first encounter.

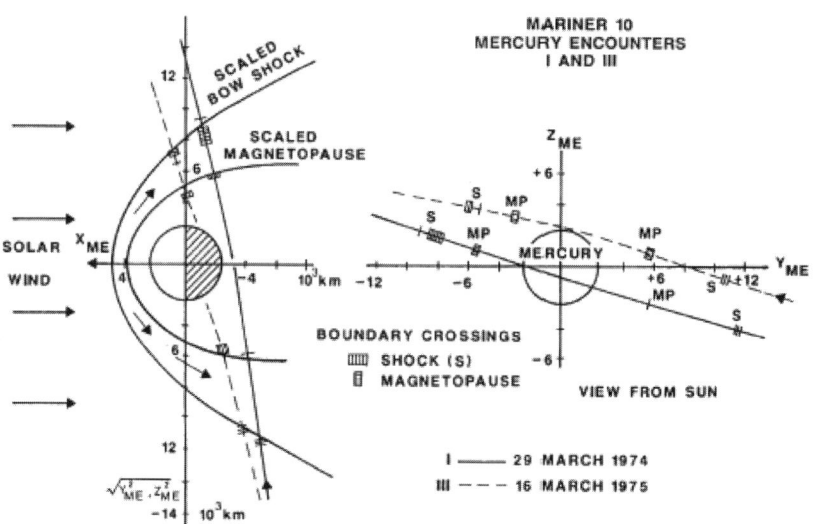

Figure 6-3. Mariner 10 Trajectories during first and third encounters in planet centered solar ecliptic coordinates as seen above the ecliptic plane (left) and from the sun (right). (Russell and Walker, 1985, copyright AGU. Reproduced by permission AGU.)

Trajectories, for Encounter I at lower latitude and for Encounter III at higher latitude, are indicated in **Figure 6-3** (Russell and Walker, 1985). In this figure, the probable locations of magnetospheric boundaries (bow shock and magnetopause) are identified.

Encounter 1 seen from a different perspective in **Figure 6-4** (Ness et al, 1975) offers additional insight by plotting average vector components in two different planes. The traces of Mercury's magnetopause and bow shock boundary are scaled for the case of a dipole moment of Mercury equal to 7 x 10^{-4} of the Earth's moment. A few of the vectors representing observations made outside the magnetopause when the spacecraft was within the magnetosheath are also shown to illustrate the sharp and distinctive change in the field direction which occurred at these boundaries. The quiet interval of the magnetosphere observations is specifically labeled and, it was noted that the magnetic field displayed a directional sense analogous to that of the Earth's magnetosphere on the near dark side. While large scale disturbances in the magnetic field after Closest Approach are evidenced in the lower diagram, the overall preservation of the magnetic field direction despite these variations is illustrated in the upper portion of the figure (the magnetic field directions remains roughly parallel to the +Z axis and positive throughout the pass).

A well defined feature in the plot of magnetic field measurements obtained during the first inbound encounter with Mercury's magnetopause is

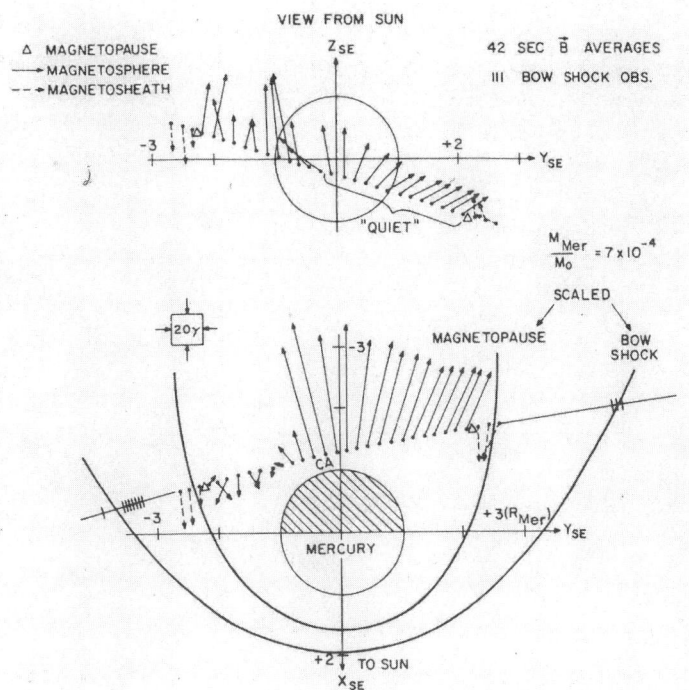

VIEW FROM SUN

△ MAGNETOPAUSE
⟶ MAGNETOSPHERE
--⟶ MAGNETOSHEATH

42 SEC $\vec{B}$ AVERAGES

III BOW SHOCK OBS.

Figure 6-4. Observed 42-second-average Magnetic Field Vectors superimposed on the trajectory of Mariner 10 in X-Z (top) and X-Y (bottom) planes during Encounter I. (Ness et al, 1975b, copyright AGU. Used with permission of AGU.) Three actual magnetopause crossings are indicated along with the (detached) bow shock transit observations. These boundaries represent a best graphical fit obtained by scaling the case of the solar wind interaction with the Earth for M = 7.0 x 10-4 of the Earth's magnetic moment.

typical of a reconnection or Flux Transfer Event (FTE) (**Figure 6-5**) (Russell and Walker, 1985). Reconnection is the merging of magnetic field lines leading to reorientation and rearrangement of particles and accompanied by release of energy, acceleration of particles, and creation of a shock wave (Kallenrode, 2004). Relative to similar phenomena at the Earth, FTEs at Mercury are of similar strength but their durations are shorter. They attain only about 6% of the size of a terrestrial FTE, although, proportionally, they have a similar scale. Also, they occur more frequently: once every minute at Mercury as compared with once every eight minutes at the Earth. Further discussion of this phenomenon can be found in Russell et al (1988).

6.3   MARINER 10 MAGNETOMETER MEASUREMENTS

The magnetometer itself recorded multiple signatures of bow shock and magnetopause crossings during the Mariner 10 mission. Detection of the

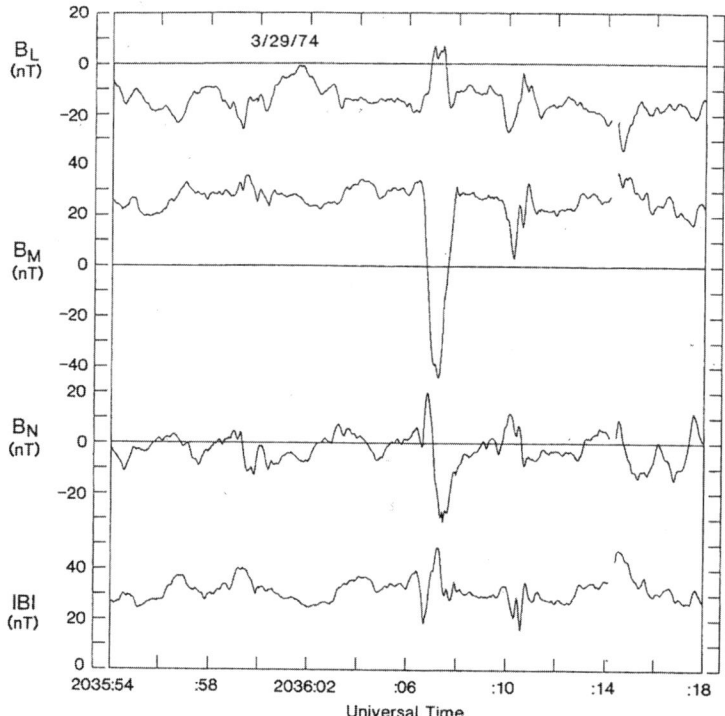

Figure 6-5. First Recorded Flux Transfer Event (FTE), a well defined feature illustrated by three components of the magnetic field near the Magnetopause on the first inbound pass (Encounter I) of Mariner 10, plotted in boundary normal coordinates. (Russell and Walker, 1985, copyright AGU. Reproduced by permission AGU.) The time resolution is 0.04 seconds. The north component points outward along the normal to the Magnetopause.

bow shock during Encounter I was accompanied by a pronounced rise in the magnetic field strength (**Figure 6-1**). Three crossings were made in a one minute interval due to the bow shock 'flapping' which was also observed with other instruments. Following this, magnetometer measurements indicated a disturbed magnetic field regime, in accord with what would be expected inside a steady state magnetosheath. About ten minutes later, a sharp boundary (the magnetopause) was traversed, as evidenced by an increase in the magnitude of the magnetic field and a decrease in the level of field fluctuations. Meanwhile, the field direction (φ) abruptly changed abruptly by 135°.

The magnetic field next increased steadily to reach a maximum value about twenty minutes later at Closest Approach. The direction of the magnetic field was then mainly parallel to the Mercury-Sun line, with a polarity sense away from the planet. A smooth, but small, variation in the orientation of the field occurred during this period. A distinct change in the previously quiet character of the magnetic field took place as indicated by large amplitude variations over a wide range of timescales. A large depression in the field occurred precipitously, followed by several other significant changes. Meanwhile, the field direction changed steadily until it pointed northwards relative to the ecliptic. Variability in the field magnitude was not matched by comparable field direction variations and, overall the general topology of the magnetic field was preserved. Minutes later, the magnetopause was crossed outbound, as evidenced by a large change in the longitudinal direction (θ) of the magnetic field. Within the outbound magnetosheath, the fields were highly variable in both direction and magnitude. Neither outbound bow shock nor magnetopause crossing are as well defined as inbound crossings. However, using data with higher resolution, the experimenters recognized that this traversal took place several times within a couple of minutes due to bow shock 'flapping'.

Determining the characteristics of the intrinsic planetary magnetic field from these data was difficult for a number of reasons. The disturbance following Closest Approach and the small size of the Mercury's magnetosphere contributed greatly to the difficulty. Even at Closest Approach, the magnetic field observations were made in regions which were not very distant from the electrical currents flowing in the magnetic tail and magnetopause. The traditional method of representing a planetary magnetic field is to utilize an expansion in spherical harmonics. In such a representation, the magnetic field is then derivable as the gradient of the scalar potential. In the case of Mercury, it was necessary to take into account the external sources of the magnetic field as well as the internal sources.

Within the constraints of the experimental sample, Ness (1979) deduced the internal dipole moment to be 5.1 ± 0.3 Tm^3 oriented at a solar ecliptic latitude of $-80° \pm 5°$ and at a longitude of $+285° \pm 10°$. This moment compares well with that deduced from the positions of the magnetopause and bow shock boundaries and with the inferred magnetic moment responsible for the deflection of the solar wind. The polarity of Mercury's dipole was found to be identical to that of the both the magnetospheres of the Earth and Mercury would be expected to have a similar response to the variable interplanetary magnetic field.

The perturbation magnetic field (B_{pert}) can be computed from the intrinsic planetary field during the magnetosphere passage. Ness (1979) accomplished this by using 42 second averages of the data, as illustrated in **Figure 6-6**. In this diagram, the magnetic field shows a predominantly southward directed orientation, and a magnitude which varies relatively smoothly from the inbound magnetopause crossing to closest approach to the outbound (dawn side) magnetopause. The sense of the field is exactly what would be expected from a magnetic tail current sheet on the night side of the planet. The orientation of the field as viewed in the X-Y plane shows a characteristic change in direction as the spacecraft passed from below the

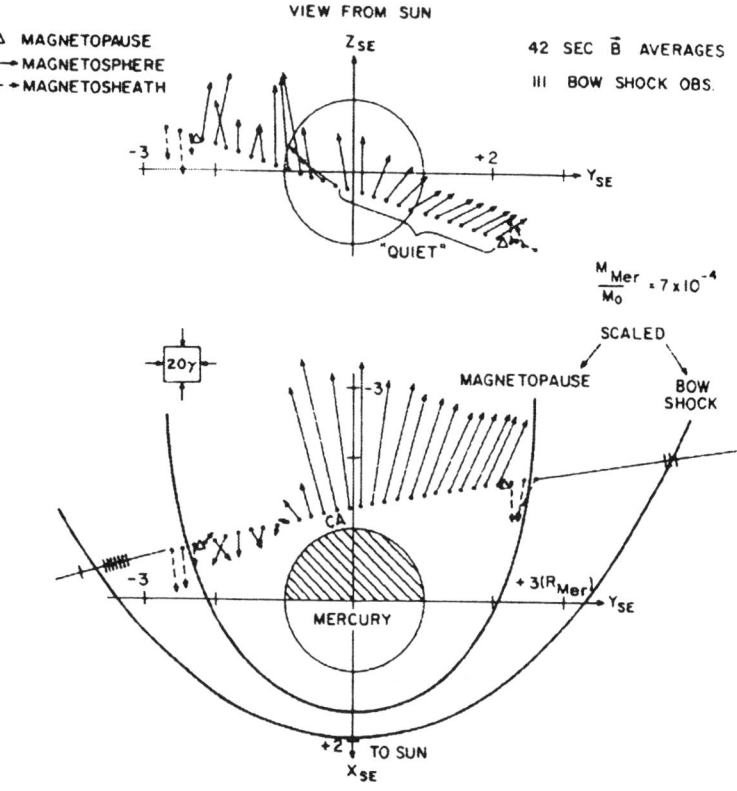

Figure 6-6. Along-Trajectory Perturbation Magnetic Field Vectors on Y-Z and X-Y solar ecliptic planes. While Bz is mainly negative, especially close to the planet, Bx-y field shows a sudden, characteristic rotation by 90 degrees as Mariner 10 crosses the neutral sheet-equatorial region of Mercury's Magnetosphere (Ness, 1979). Courtesy of Norman Ness.

magnetic equatorial line to above it, (This change in the perturbation field direction is also identical to what would be expected in the case of the terrestrial magnetosphere if a spacecraft carrying a magnetometer passed from the southward lobe of the magnetic tail to the northward lobe, near X = 8 to 12 R_E.) The change in field direction and drop in field magnitude at this time (compare with **Figure 6-1**) suggest that a strong current sheet was crossed which reversed the polarity of the X component of the magnetic field. This decrease is interpreted to be diamagnetic and to have corresponded to spacecraft entry (from below) into a high-beta, central, plasma sheet, analogous to the region separating the two lobes in the Earth's magnetotail. Studies of this region by Russell et al (1988) support the interpretation that Mariner 10 indeed transited a cross-tail current sheet after Closest Approach, rather than that it traversed a temporal variation in the background field. The abrupt decrease and recovery indicated in **Figure 6-1** may have been due to an intensification of the neutral sheet current as the tail sheet increases and/or to a motion of the edge of the neutral sheet closer to the planet (Ness et al, 1975).

The existence of a modest intrinsic magnetic field at Mercury sufficient to deflect the solar wind and of an imbedded neutral sheet, leads to the conclusion that Mercury should have a magnetotail. The expected tail-like character of the field is evidenced by the magnetic measurements from the inbound trajectory, where $|Bx| >> |By|$ and $|Bz|$. Measurements of the magnetic field just after the inbound magnetopause crossing, suggest (Ness, 1979) that for Mercury the tail field is 30-40 γ with a radius of ~ 2.0 to 2.6 R_M. The assumed magnetic dipole of 5.1×10^{12} Tm^3 leads to a polar cap co-latitude (θ_{PC}) value of 17 to 26°, approximately two times that of the Earth's. Direct entry of solar wind plasma into the magnetosphere can, thus, take place more efficiently at Mercury where it is estimated that the penetration of plasma through the dayside cusps can extend down to latitudes of 60° (Kabin et al, 2000).

During Encounter III, these high latitude magnetic field measurements were far quieter than those that characterized Encounter I, particularly after Closest Approach (**Figure 6-2**). The interplanetary magnetic field was directed northward both before and after this encounter. Such a configuration is unfavorable (Siscoe et al, 1975) for dayside reconnection (with the associated injection of energy into the magnetosphere). This can explain why very smooth profiles were recorded during this transit. When corrected for the different distances of Closest Approach during the two encounters, the polar magnetic fields measured during III were revealed to be about twice as large as those measured along the low latitude Encounter I trajectory. The strong, well ordered, magnetic fields recorded proved that Mercury has a significant intrinsic magnetic field that is primarily bipolar and interacts with the solar wind to produce a magnetosphere. Against the background of experimental uncertainties, the magnitude of the bipolar moment was deduced to be between 300-500 nTR^3_M with a tilt relative to the planetary rotation axis of about 10° (Connerney and Ness, 1988).

The Encounter I and III data are complementary. The magnetic field measurements made during Encounter I, while not as useful for deriving the intrinsic magnetic field of the planet, nevertheless provided important basis for understanding magnetospheric current systems, which are driven by the solar wind interaction.

6.4   ORIGIN OF MERCURY'S MAGNETIC FIELD

The origin of Mercury's internal magnetic field is still controversial. The two most likely explanations are either intrinsic magnetization of sub-Curie point material in the outer layers of the planet or the presence of an active dynamo in the interior (Connerney et al, 2001). Gravity field measurements indicate that Mercury is very likely to possess a large iron core (Anderson et al, 1987; Schubert et al, 1988), supporting the latter hypothesis. Although the pure iron core of a planet the size of Mercury should have cooled to the solidus point prior to the present time (Solomon, 1976; Peale, 1976), the mixing of a small amount of an alloying element such as sulfur or hydrogen would lower the melting point and prevent core freezing (Schubert et al, 1988; Okuchi, 1997). The details of Mercury's internal magnetic field are currently a subject of debate and the only consensus seems to be that a centered dipole is not a good approximation (Blomberg and Comnock, 2004). This topic is discussed in detail in Chapter 3.

6.5   MARINER 10 PLASMA OBSERVATIONS

The Mariner 10 payload included a plasma characterization instrument. Its operation was hampered by a deployment failure that precluded its ion measurement capability. However, the electron detecting portion of the instrument showed nominal operation. Counting rates recorded by the Plasma Electron Spectrometer in selected energy channels (Ogilvie et al, 1974, 1977) during Encounters I and III are presented in **Figure 6-7** (Ness, 1979).

The locations of strong aligned currents associated with field line crossings, including the individual bow shock and magnetopause boundaries identified on the basis of plasma data show good correspondence with those identified by magnetic field measurements. Further, plasma speed and density parameters derived from the electron data are consistent with regard to bow shock jump conditions and pressure balance across the magnetopause (Ogilvie et al, 1977, Slavin and Holzer, 1979). Within Mercury's magnetosphere, the measured plasma density was higher than that observed at the Earth by a factor equivalent to the ratio of the external solar wind density at the orbits of the two planets.

In the course of Encounter I, plasma sheet type distributions were recorded that showed an increase in plasma temperature during the disturbed conditions characterizing the outbound part of the pass. During Encounter III, in the high latitude magnetosphere, the instrument recorded the 'horns' of a cool, quiet time, plasma sheet (Ness, et al, 1976; Ogilvie et al, 1977).

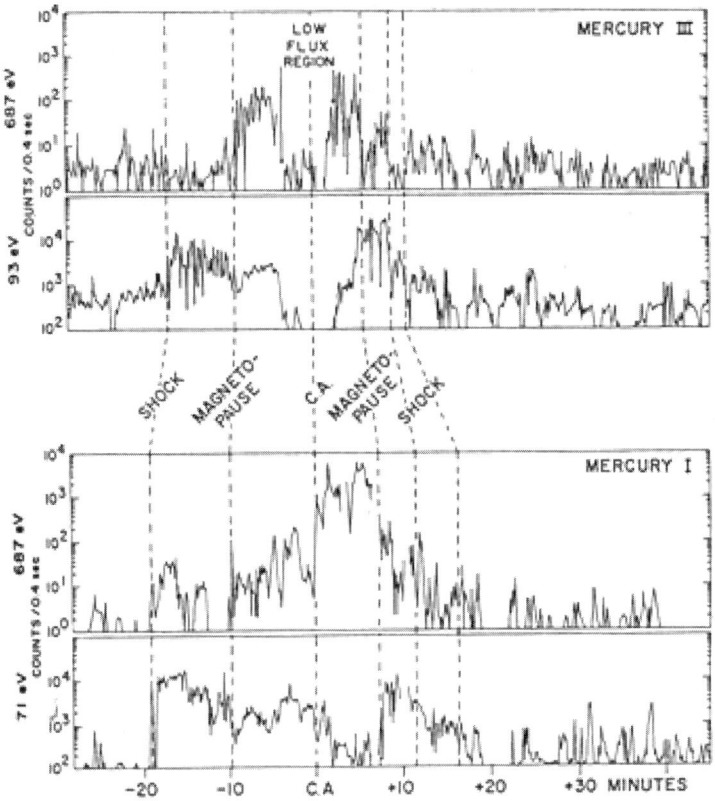

Figure 6-7. Along-Trajectory Electron Spectrometer Measurements (from two channels) on Mariner 10 during Encounter 1 (top) and Encounter 3 (bottom). (Ness, 1979) Courtesy of Norman Ness.

As shown in **Figure 6-7** (Ness, 1979), entrance into the magnetosphere was marked, during both encounters, by a rise in the high energy flux and a drop in the low energy flux. The electrons recorded were similar to those found in the plasma sheet boundary layer of the Earth. No cold dense plasmas indicative of ionospheric or plasmaspheric particles were encountered. This observation is consistent with the relatively larger (8x) scale of the planet Mercury relative to the Earth in relation to their respective magnetospheres. The plasma sheet densities at Mercury were higher by a factor of five than those measured at the Earth (which is close to the ratio of the solar wind density at 0.5 to 1.0 AU). This supports the hypothesis that, in the absence of a significant ionosphere, the solar wind is the primary source of this population (Ogilvie et al, 1977).

Figure 6-8 (Ness, 1979) depicts the locations of the various regimes identified in the plasma electron data along the Mariner 10 trajectories as seen from the Sun during Encounters I and III. Also indicated is the location

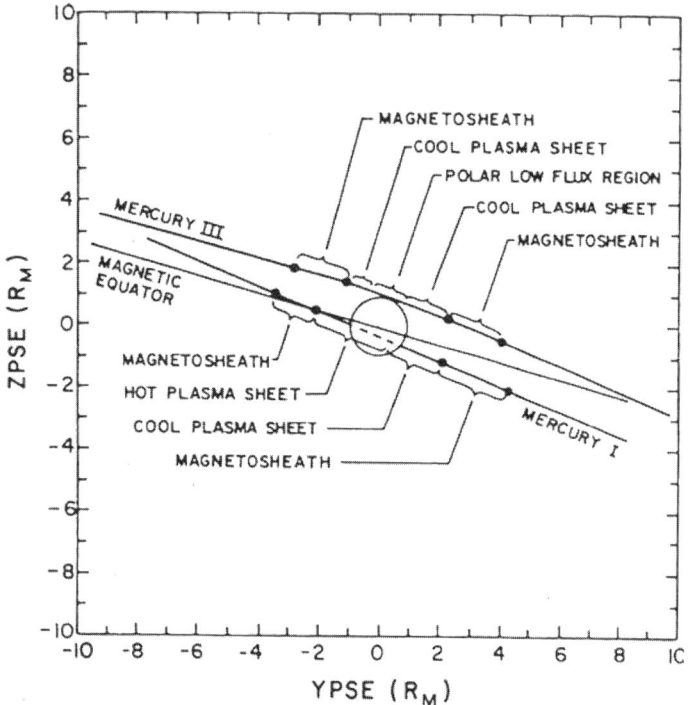

Figure 6-8. Along-Trajectory Plasma Regimes of Mariner 10 during Encounter I and Encounter III as observed from the sun (Ness, 1979). The plane shown is the Y-Z planetary solar ecliptic projection. Plasma electron regimes are indicated with the magnetic equator or current sheet for reference. Courtesy of Norman Ness.

of the magnetic equator (or current sheet). The hot plasma sheet mentioned was observed during the disturbed outbound portion of Encounter I and the cool plasma sheet during the inbound quiet period.

6.6   MARINER 10 ULF OBSERVATIONS

Event propagation in magnetospheres can be measured with antennas which detect frequency variation. On Mariner 10, this instrument was called the UltraLowFrequency wave detector. Waves may be electrostatic, where only the E field fluctuates while the magnetic field remains static, or electromagnetic, where both E and B fields oscillate; electron waves involve higher frequency oscillations of the smaller mass and inertia electrons with ions creating an unchanging background, while ion waves involve lower frequency oscillation of higher mass and inertia ions and electrons combined (Kallenrode, 2004). How effectively waves are propagated depends on the orientation of wave vectors relative to the surrounding magnetic and electric fields (Kallenrode, 2004). Affecting the entire magnetosphere, pulsating

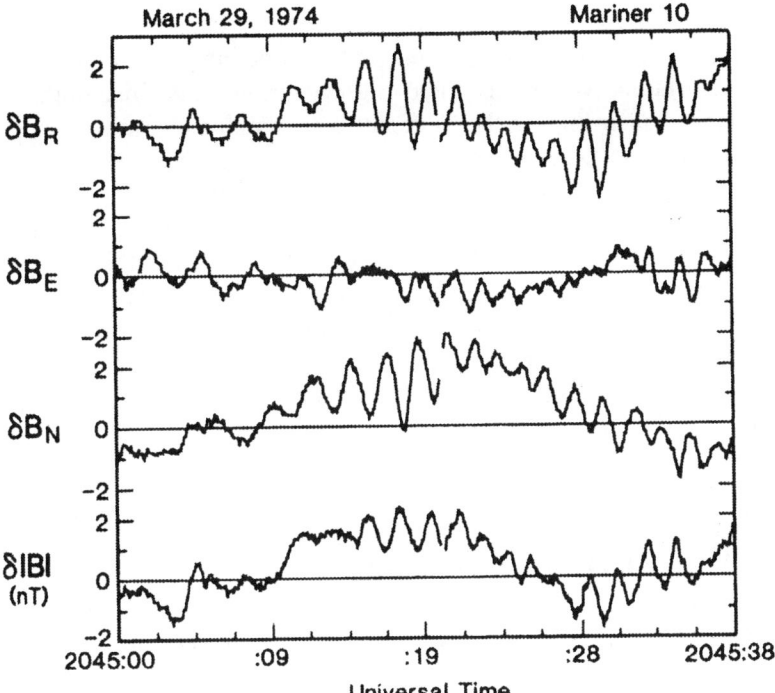

Figure 6-9. Ultra Low Frequency (ULF) waves (standing Alfven waves assumed to be fourth harmonic) recorded in Mercury's magnetosphere aboard Mariner 10 during Encounter I (Russell, 1989, copyright AGU. Used with permission of AGU.) Mean field during the interval has been removed. Coordinate system is radial from the planet, east and north. Units on the y-axis in nT.

waves result from geomagnetic activity most influenced by the sun's magnetic cycles (Campbell, 1997).

In the terrestrial magnetosphere (damped) standing waves are frequently generated by a Kelvin-Helmholtz instability at the flanks of the magnetosphere as the solar wind streams past. The presence of standing Alfven waves in Mercury's magnetosphere has been reported by Russell (1989). This These waves are often referred to as field line resonances since they involve a large scale, fluctuating, motion along the entire length of a set of magnetic field lines.

Narrow band ULF waves recorded as shown in **Figure 6-9** (Russell, 1989) have a period of about 2s, and a bounce period of about 8s, and are primarily transverse to the ambient field. The polarization of these oscillations is essentially in the magnetic meridian direction. Russell (1989) suggested that this signal was caused by a standing Alfven guided by the magnetospheric field lines.

Southwood (1997) noted that the discovery of ULF signals at Mercury provides insight on the closing of magnetospheric currents in the vicinity of planetary surfaces. Apparently, field aligned currents can close in the mantle of the planet, rather than in the more perfectly conducting, and deeper, core. Depending on the conducting properties of the planet and its immediate environment, the waves will either be reflected above, at, or below the surface. Also, depending on whether the conductance at the reflection boundary is greater than, or less than, the conductance of the waveguide, either the magnetic field or the electric field of the wave will change phase when it is reflected. Since the electric field was not measured by Mariner 10, no firm conclusion regarding the nature of the waves observed can presently be drawn. Field Aligned Currents will be discussed in greater detail in Section 6.9.

6.7   MAGNETOSPHERE STRUCTURE

Many workers have modeled the dimensions of the dayside magnetosphere based on the Mariner 10 magnetopause and bow shock transit measurements (Ness et al, 1974; Ogilvie et al, 1977; Russell, 1977; and Slavin and Holzer, 1979a, 1979b). A straightforward way to understand Mercury's magnetosphere is to scale the Earth's magnetosphere to be equivalent to Mercury's. This is done by decreasing the radius of Earth's magnetosphere by a factor of 8 relative to the radius of the planet and then decreasing the magnetosphere's strength by a factor of 1000 (Ogilvie et al, 1977).

Figure 6-10 (Russell et al, 1988) represents Mercury's magnetosphere scaled to occupy the same volume as the terrestrial magnetosphere. The plasma sheet is seen to almost touch the planet's surface near midnight, while the polar cap field, consisting of field lines entering the magnetotail, extends to very low planetary latitudes on the night-side. Mercury occupies the location of the plasmasphere and the region of the most intense Van Allen Belts. Based on size alone, these features of the terrestrial magnetosphere should be absent at Mercury.

Another representation of Mercury's magnetosphere can be seen in **Figure 6-11** (Russell et al, 1988) again scaled so that it occupies the same volume as the terrestrial magnetosphere. The solid lines with arrows show the motion of low-energy (or cold) plasma that drifts across magnetic field lines all perpendicular to the plane of the diagram due to the electric field applied to the magnetosphere by the solar wind. The strength of the electric field is proportional to the strength of the coupling between the solar wind

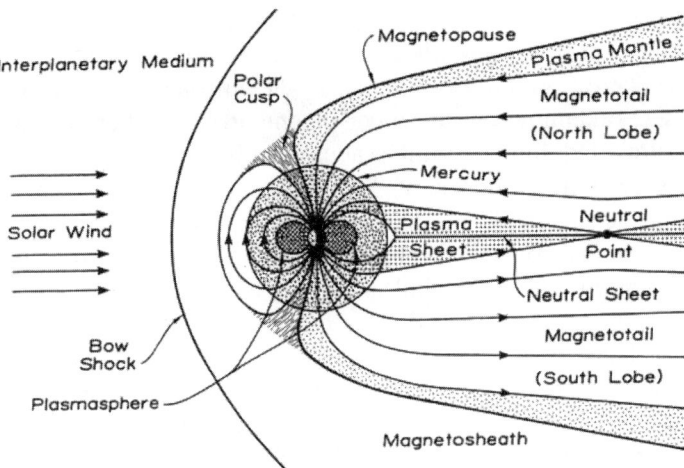

Figure 6-10. Mercury's magnetosphere scaled to Earth's, Side View. (Russell et al, 1988, The Magnetosphere of Mercury, in Mercury, Copyright 1988, The Arizona Board of Regents. Reprinted by permission of the University of Arizona Press.) In this noon-midnight depiction, the planet occupies a large fraction of the Earth's inner magnetosphere, including the plasmasphere and the most intense Van Allen Belts.

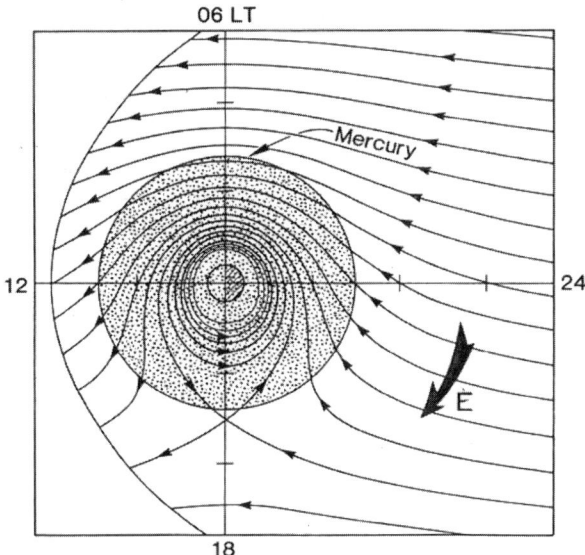

Figure 6-11. Mercury's Magnetosphere scaled to Earth's, Top View. (Russell et al, 1988, The Magnetosphere of Mercury, in Mercury, Copyright 1988, The Arizona Board of Regents. Reprinted by permission of the University of Arizona Press.) In this equatorial plane projection, the planet blocks the entire region of co-rotating plasma. Such a region would not be expected at Mercury, even if the planet were smaller relative to its magnetosphere, due to the planet's lack of dense atmosphere and ionosphere and its slow rotation resulting in streamlines much straighter than those shown.

and the magnetosphere. Earth's highly conducting ionosphere and dense atmosphere result in a corotating plasma region of the electric field called the plasmasphere (Encrenaz et al, 2003). The co-rotating plasma produces a set of closed streamlines in the inner magnetosphere where plasma can, in principle, circle indefinitely while gradually building up its density from low altitude sources in the ionosphere. This process leads to the formation of the Earth's high density, cold plasma, region (the plasmasphere). Outside this regime, plasma is convected out through the magnetopause due to the coupling between the solar wind and the magnetosphere. Mercury rotates more slowly than the Earth and does not have a substantial ionosphere or (Chapter 5). Thus, even if Mercury itself did not occupy such a large portion of its magnetosphere, it would not be expected to have a plasmasphere. Also, the flow lines would be different for Mercury, far straighter than those drawn in **Figure 6-11**. Any cold or low-energy plasma entering into the equatorial region would be swept out through the magnetopause, Any ionized component would be swept out into the solar wind.

Figure 6-12 (Slavin, 2004) is a cutaway picture of Mercury's magnetosphere scaled to the Earth's. In this scenario, the mean distance (1.5 R_M where $1 R_M$ = 2439 km) from the center of Mercury to the sub-solar magnetopause maps to about 12 R_E at the Earth (where 1 R_E = 6378 km). On this scale, the surface of Mercury is located at a distance just beyond geosynchronous orbit on the Earth. Also, the observed near-tail Hermean diameter of ~ 5 R_M scales to about 40 R_E ,(close to the mean diameter of the Earth's tail). A characteristic length scale of any magnetosphere is the distance from the center of the planet to the sub-solar point of the magnetopause. Based on Mariner 10 magnetic field measurements, this length was estimated by Ness (1979) to be 1.45 ± 0.15 R_M. However, in all likelihood, this length would typically be somewhat larger (1.8 ± 02 R_M) due to solar wind variability and to the highly eccentric orbit of Mercury (Siscoe and Christopher, 1975).

6.8   MAGNETOPAUSE STRUCTURE

Mariner 10 data indicated that the Mercurian magnetopause had a significant aberration in the ecliptic plane as well as equator-pole asymmetries. Ness et al (1975) computed the shape of the magnetopause for the case where the solar wind was incident on a Mercury-centered magnetic dipole orthogonal to the solar wind flow. The theoretical position of the bow shock was determined for the case of aligned flow, in which the upstream magnetic field and solar wind velocity were deemed to be parallel. A sonic Mach Number 10 and an Alfven Mach Number 20 at the subsolar point were

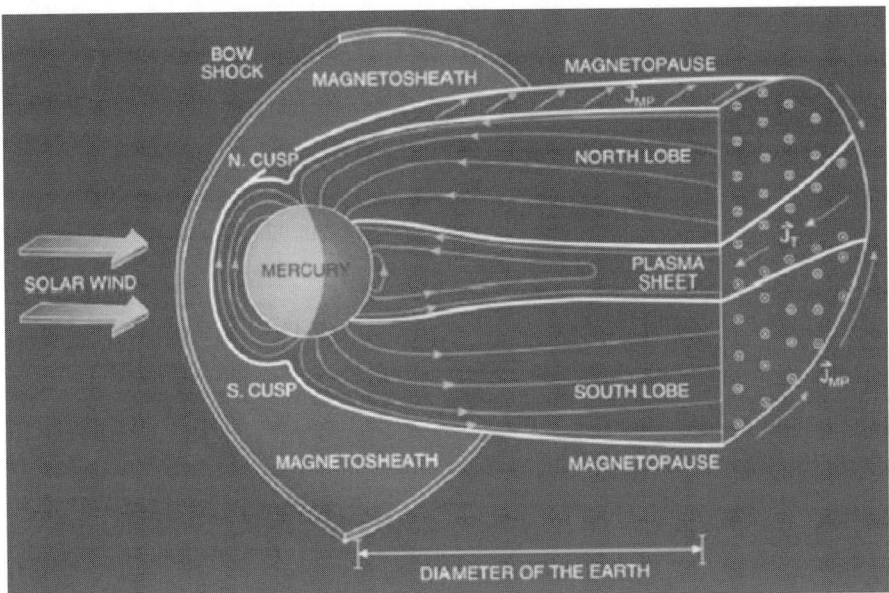

Figure 6-12. Schematic view of Mercury's Magnetosphere obtained by superimposing an image of Mercury increased in size by a factor of 8 onto Earth's magnetosphere. (Reprinted from Slavin, 2004 with permission of Elsevier)

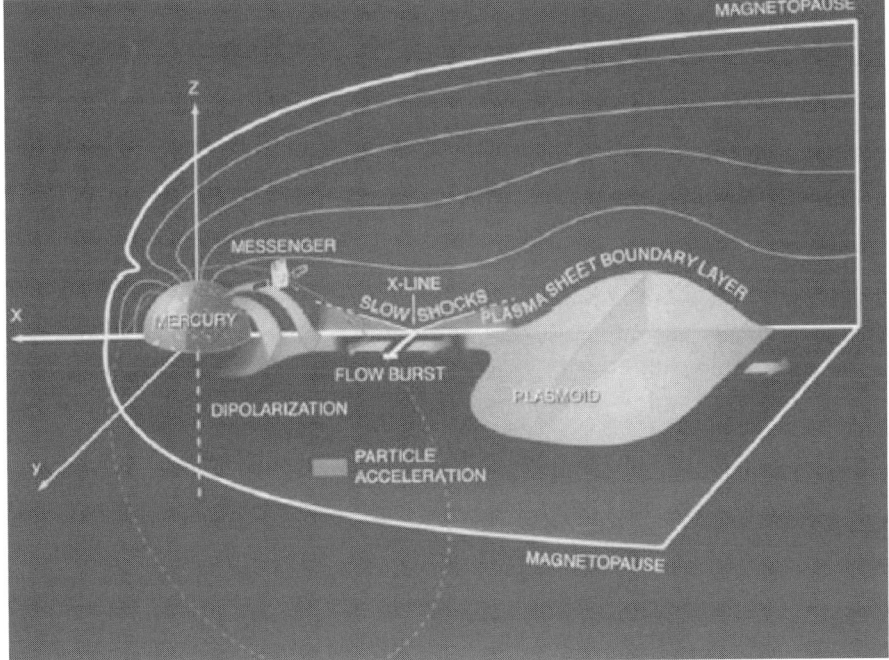

Figure 6-13. Schematic growth phase of a substorm Expansion at Mercury based on the Near Earth Neutral Line (NENL) model (Reprinted from Slavin, 2004, with permission of Elsevier).

adopted, based on combining the solar wind conditions experienced during Encounters I and III. A somewhat better correspondence with the observations was later achieved by taking into account the effect of aberration due to planetary heliocentric motion (Ness, 1979). This involved considering the solar wind flow to impinge on Mercury from a location at five degrees to the east of the solar direction, rather than parallel to the Mercury-Sun line. Cylindrical symmetry in the shape of the bow shock and magnetopause boundaries about the direction of the solar wind flow was, in each case, assumed.

Russell et al (1988) stressed that extrapolating from the Mariner 10 bow shock and magnetopause boundary crossings to the subsolar region is fraught with uncertainties due to lack of knowledge of crucial parameters including boundary shapes, solar wind direction, and the influence of boundary flapping. Nevertheless, on carrying out such an extrapolation using an aberration angle of 8^0 rather than 5^0, these authors obtained a subsolar magnetopause distance of $1.35 \pm 0.2R_M$ and a subsolar shock stand-off distance of $1.9 \pm 0.2R_M$. The ratio of these quantities is 1.41 ± 0.3 (which may be compared with the corresponding value of 1.3 at the Earth). However, the ratio derived is too inherently uncertain to allow it to be utilized to make deductions concerning the shape of the magnetopause.

The assumed position of the subsolar magnetopause can be used to determine the magnetic moment of Mercury if the dynamic pressure of the solar wind is known and if the magnetosphere contains a low-beta plasma. In this regard, Russell et al (1988) selected a magnetospheric shape factor appropriate for a gas dynamic interaction with a ratio of specific heats of 5/3. They ignored plasma and tail current effects and used representative values for the density and velocity of the solar wind measured by the electron instrument (Slavin and Holzer, 1979). They thus derived a magnetic moment for Mercury of 1.5×10^{12} Tm^3. By utilizing, thereafter, different measured solar wind measurements and including various corrections discussed by Slavin and Holzer (1979) other possible values for the magnetic moment were estimated and found, overall, to lie between 1.5×10^{12} and $6.0 \pm 2 \times 10^{12}$ Tm^3. Some authors (e.g., Whang, 1977) attempted to derive multi-pole moments for Mercury. This genre of quadrupole and octupole modeling studies was criticized by Ness (1979) who argued that the derived moments are due to spatial harmonic aliasing due to insufficient planetary coverage by Mariner 10 to allow the higher moments to be calculated.

Siscoe and Christopher (1975) and Goldstein et al (1981) examined the statistics of the location of the subsolar magnetopause of Mercury in an attempt to determine how often the solar wind strikes the Hermean surface. Although these authors used slightly different initial assumptions, both concluded that, such strikes occur but seldom. For example, for a magnetic moment of 2.5×10^{12} Tm^3, Goldstein et al (1981) estimated that the probability of a strike varies from 6.1×10^{-5} at aphelion with no tangential stress to 6.6 x 10^{-2} at perihelion with no tangential stress. In a related discussion, Russell et al (1988) noted that if a subsolar radius of 1.2 R_M is adopted, the associated planetary magnetic moment will be associatively lower and the predicted solar wind strikes more frequent. Dayside reconnection would significantly increase the chance for the solar wind to impact with the Hermean surface (Slavin and Holzer, 1979).

An important effect resulting from the high electrical conductivity of Mercury's core and mantle was indicated by Suess and Goldstein (1979) and by Hood and Schubert (1979): rapid solar wind variations will induce currents in these highly conducting regions in a direction that acts to oppose magnetopause compression.

Conversely, at Earth, magnetic field lines are transported from the dayside magnetopause to the magnetotail through the process of reconnection (Russell and McPherron, 1973). Characteristic signatures of non-continuous reconnection are easily identified in terrestrial magnetic field data displayed in a co-ordinate system that is oriented to the plane of the magnetopause.

6.9   MAGNETOSPHERE DYNAMICS

Mercury's magnetosphere is the most dynamic in the solar system. The implied substorm time scales at Mercury are ~33 times faster than at Earth (Siscoe et al, 1975; Christon, 1987). The brevity of magnetospheric phenomena at Mercury can be typified in part by the 1 to 2 minute duration of the substorm (Ness et al, 1975; Ogilvie et al, 1974; Simpson et al, 1974), significant electron flux increases in less than one second, and plasma electron flux variations in the magnetopause in less than ten seconds (Christon, 1987).

The first and only measurements of energetic particles (E>170 keV) in Mercury's magnetosphere were made by the energetic particle instrument on Mariner 10. Energetic particle measurements made during Encounter I are compared with contemporaneous plasma and magnetic field data in **Figure 6-14**. The bottom half of the figure presents a magnetospheric model produced by Whang and Ness (1975) to explain the particle data. Several enhancements of the energetic electrons are individually labeled A, B, B', C, D and D', following Eraker and Simpson, 1986. Consideration of the velocity of the spacecraft, the electron gyroradius, the count rate increases and the magnetic field data lead Simpson et al, 1974 to the conclusion that these particle bursts were transient events and not spatial structures. Also, it was shown by Christon et al (1987) that all of the particle events but A (which occurred before Closest Approach) occurred during the outbound leg of the pass, event C straddled the outbound magnetopause crossing, while D

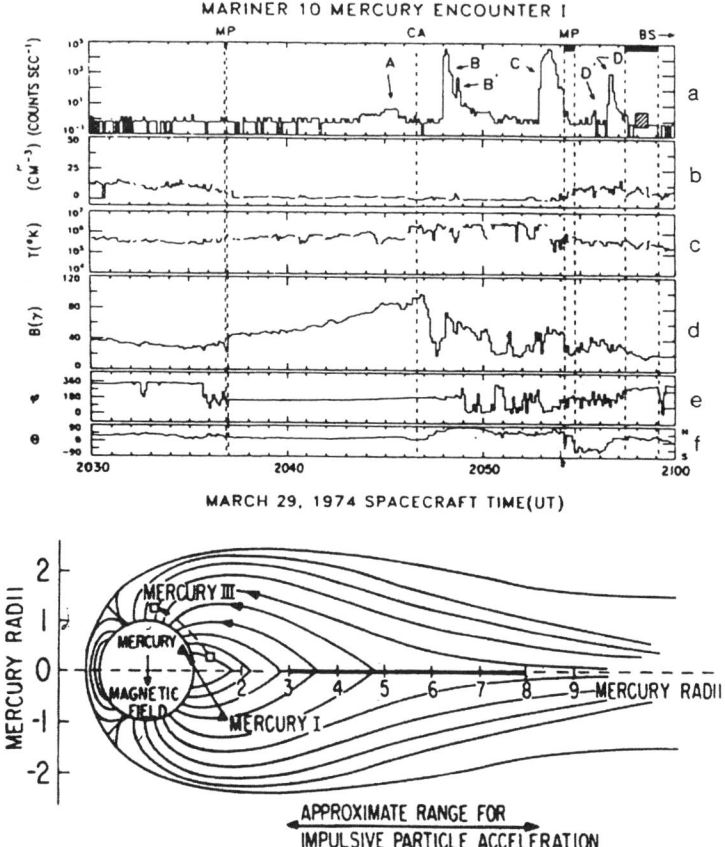

Figure 6-14. Particle Burst Measurements and Model. Upper half of figure (Ness, 1979, used with his permission) presents counting rates recorded during Encounter I by an instrument designed to measure electrons with E>170 keV. Important burst events are labeled A-D. Panels 2-6 illustrate relationship of particle bursts to plasma and magnetic field activity, showing, respectively, plasma density, electron temperature, magnetic field magnitude (B), azimuth (γ) and inclination (θ). Lower half of figure shows a magnetospheric model developed by Whang and Ness (1975, NASA document) to explain particle data.

was observed in the dawn-side magnetosheath. No other energetic particle enhancements were observed close to Mercury.

A preliminary analysis of the energetic particle data by Simpson et al (1974) indicated that protons with energies up to approximately 550 keV, and electrons with energies up to about 300 keV, were simultaneously present in the Hermean magnetosphere. Later, Armstrong et al (1975) and Christon et al (1979) reconsidered the data. Both groups demonstrated that these records cannot be interpreted uniquely but are consistent with either simultaneous fluxes of energetic electrons and protons or intense fluxes of electrons only (the latter exhibiting very steep differential energy spectra). They questioned the previous claim that protons apparently present could be explained as an effect of electron pile-up. Thus, the origin of particle signatures recorded during Encounter I, whether the result of proton and electron bursts or electron bursts alone, remains unresolved. While the latter interpretation is generally adopted in the literature for interpretative purposes, the possibility that protons are present at Mercury cannot be excluded until suitable experiments are performed.

Siscoe et al (1975) and Ogilvie et al (1977) noted that when Mariner 10 entered the near-tail below the plasma sheet on the dusk side during Encounter I, the sheath field was northward. At that time, the field inside the magnetopause was tail-like and relatively quiet. Shortly after Closest Approach, the magnetic field strength decreased rapidly and the field inclination significantly increased, thereby indicating a transition from a tail-like to a dipole-like field orientation. The large changes in the magnetic field strength occurred in the same time interval as the large energetic particle bursts B, B' and C. Event A, which occurred before Closest Approach, appears to have had a (weak) magnetic field change associated with it. The magnetic field was aligned strongly southward at the time when Mariner 10 exited the magnetotail. On the basis of these observations, Siscoe et al (1975) suggested that, when Mariner 10 was about half way across the tail, the interplanetary magnetic field (IMF) changed from northward to southward, thereby (in analogy with what can be observed at the Earth), initiating strong, sunward, plasma sheet convection and a series of substorms. Scaling arguments were used to demonstrate that substorm time scales should be of the order of 1 to 2 minutes at Mercury as compared with 30-60 minutes at the Earth (see also Slavin and Holzer, 1979).

This scenario originated by Whang and Ness (1975) was further developed by Eraker and Simpson (1986) who suggested that the substorms inferred to be present in Mercury's magnetotail resulted from reconnection events at 3-6 R_M associated with the acceleration of energetic electrons that gave rise to the enhancements recorded. The model developed by these authors shows the likely region of particle acceleration.

Siscoe and Christopher (1975) pointed out that the timescale of magnetospheric processes should be proportional to the convection time (Tc) required to recycle the flux in the tail (F_T) across the magnetosphere at the electrostatic potential (φ_C) where $Tc = F_T\ \varphi_C$. If the sunward position of the magnetopause (mp) is R_{MP} and the magnetic field in the tail region is B_T, the total magnetic flux can be approximated by (*Equation 6-1*):

$$F_T = 2\pi R^2_{MP}\, B_T$$

The cross tail electrostatic potential (in the solar wind, sw) will be $V_{SW}B_{SW}R_{MP}$. With substitution, this yields (*Equation 6-2*):

$$T_C = 2\pi B_T \, R_{MP}/B_{SW}R_{MP}$$

On substituting representative values for the relevant magnetospheric parameters, the average values of Tc obtained respectively for the Earth and Mercury were ~ 1^h and 2^m respectively.

According to the Near Earth Neutral Line (NENL) theory of substorm activity discussed by Baker et al (1996) and Slavin et al (2002) dynamic features may be present at Mercury at the time of substorm activity. These are shown in **Figure 6-13** (Slavin, 2004). Mercury and its environment are here scaled to the size of the terrestrial magnetosphere. The neutral line (located on this scale at 2 to 1.5 R_M), is the site where lobe magnetic field energy is dissipated, producing high speed, plasma flows, plasma heating and energetic particle acceleration. According to this scenario, jets of plasma emanate from the x-line with speeds comparable to the Alfven speed in the flux tubes undergoing reconnection. In the terrestrial magnetosphere such sunward and anti-sunward flows are called respectively 'bursty bulk flows' and 'post-plasmoid plasma sheet flows' (Angelopoulos et al, 1992; Richardson et al, 1987). The anti-sunward flows slow down as they compress the dipolar magnetic field close to the planet and, thereby, create a high pressure region that is thought to drive a strong, field aligned, current, called a substorm current wedge, into and out of the ionosphere (Shiokawa et al, 1998). In the downstream tail, moving structures called plasmoids, illustrated in **Figure 6-13**, would, according to a scaled comparison with the terrestrial case, be expected to grow to attain a diameter of several Mercury radii.

Detailed (0.04s) energetic particle and magnetic field observations recorded during the B-B' event are presented in **Figure 6-15** (Christon, 1987). Typical individual energetic particle events have a duration of several seconds (Eraker and Simpson, 1986). Less than a minute after the spacecraft entered the plasma sheet (see **Figure 6-1**), there was a rapid increase in the Bz field component. This sudden increase and subsequent quasi-periodic increases are nearly coincident with enhancements in the flux of > 35 keV electrons (Christon, 1987; Baker et al, 1986; Simpson et al, 1974). This energetic particle event and other lesser events recorded later in the outbound pass, were interpreted to provide evidence of substorm activity by both the early investigators and subsequent workers (Eraker and Simpson, 1986, Baker et al, 1986 and Siscoe et al, 1975). An analysis by Christon, 1987 of both the magnetic field and energetic particle signatures instanced strong similarities with substorm activity commonly recorded at the Earth. In particular, the sudden change in the Bz component observed in the Mercury magnetic measurements is frequently observed at the Earth where it is termed a magnetic field depolarization event. Such events are closely associated with energetic particle injections and with field aligned currents.

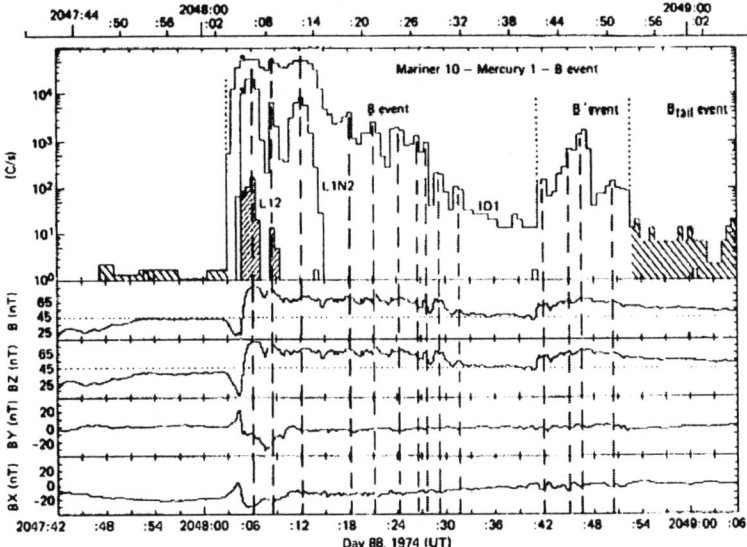

Figure 6-15. High Time Resolution Measurements of Particle Burst. (0.6 second) energetic electron (>35 keV) count rates (top) and magnetic field (0.04 second) measurements (bottom) (Reprinted from Christon, 1987, with permission from Elsevier). Vertical dashed lines mark rise times of larger energetic particle events.

Baker et al (1986) and Christon (1987) suggested that each of the particle events recorded by Mariner 10 was due to an individual substorm and the formation of a near Mercury x-line in the tail. **Figure 6-16** (Baker et al, 1986) illustrates particle acceleration at a near Mercury neutral line and the subsequent injection of energized electrons into the dipolar magnetic fields of the inner magnetosphere. Once there, particles drift eastward. Baker et al (1986) estimated that the drift time is 6 to 8 seconds for ~300 keV electrons (assuming a closed path). Events at the dawn magnetopause and in the magnetosheath were attributed to electrons injected onto drift paths that intersected the dawn-side magnetopause.

Modeling of energetic protons originating in a 1 R_M wide region located 2 to 3 R_M downstream from Mercury, indicates that, for particles with energies ≤ 10 keV, the E x B drift dominates gradient and curvature drifts so that these protons intercept the surface of Mercury near to midnight (Lukyanov et al, 2001). At energies of 30-50 keV, the curvature and gradient drifts become more important so that a partial ring forms, with many ions hitting the surface of the planet on the day side or escaping through the dawn side magnetopause. Above 50 keV, nearly all the protons encounter the planetary surface or the magnetopause within just a few seconds.

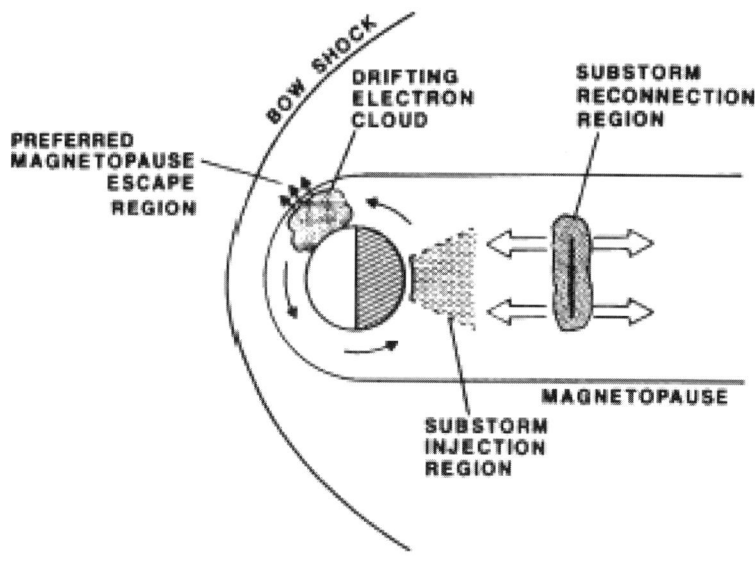

Figure 6-16. Schematic View of Energetic Particle Acceleration at a near-Mercury neutral line and subsequent injection into the inner magnetosphere (after Baker et al, 1986). Reproduced by permission of the American Geophysical Union.

A sudden change in the magnetic tail field from stretched out to a more dipolar configuration is known to be accompanied in the Earth's magnetosphere by the injection of large fluxes of energetic ions (in the range 10-100 keV) into the trapped radiation belts (Hamilton et al, 1988). A significant portion of this injected particle population appears as field aligned beams (Quinn and McIlwain, 1979). An electric field is induced during the reconfiguration of the magnetic field. The guiding centers of the charged particles will follow an E x B drift motion such that the particles will co-move with the magnetic field line. Particles which have drifted across the equatorial plane will be subject to acceleration by the electric field (E) parallel to the magnetic field (Quinn and Southwood, 1982). Thus, an ion beam with a narrow pitch angle distribution will be produced following the depolarization event (Chapman and Cowley, 1984, Lewis et al,, 1990). It was estimated by Ip (1987) that, at Mercury, protons and He^{++} ions could be accelerated in this way to a maximum energy of 13-14 keV. Similarly, sodium ions (created by photo-ionization and/or electron impact) could, during a prolonged period of intensified cross tail electric fields, be accelerated to several tens of keV. Because of their large gyroradii, such ions could strike Mercury's surface in many different places (i.e. they would not be confined to the auroral zones). During short duration (10s) reconfiguration events, these particles will have limited acceleration, due to their high mass to charge ratio.

6.10   SUBSTORM ACTIVITY

Mariner 10 provided the first evidence of substorm activity at Mercury as illustrated by magnetic field variations following depolarization events during Encounter 1 **(Figures 6-4 and 6-5)** when the magnetic field suddenly changed from a tail aligned orientation to one pointing more to northward while a strong enhancements in the fluxes of > 35 keV electrons (Slavin et al, 1997).   This event was interpreted to provide evidence of substorm activity (Siscoe et al, 1975, Baker et al, 1986, Eraker and Simpson, 1986, Christon, 1987).

Substorms at the Earth, correspond to intervals of greatly enhanced magnetospheric convection, involving the dissipation of large amounts of magnetic energy stored in the tail lobes through (a) the acceleration of charged particles (b) the driving of intense field aligned currents (FACs) linking the tail and ionosphere and (c) the generation of strong bulk flows and the production of heating in the central plasma sheet. The time necessary to recycle all of the flux stored in the tail lobes at the Earth under typical disturbed conditions is about one hour whereas, at Mercury, the corresponding time was estimated by Siscoe et al (1975) to be approximately one minute. This is in close agreement with the duration of the, representative, observed, energetic particle enhancement presented in **Figure 6-6**. At the Earth, the duration of a substorm is influenced by the effect of 'line-tying'. In this process, ions convecting in response to the electric field the magnetosphere attempts to impose on the ionosphere, experience a net drag force as a result of collisions with thermospheric neutral species. These latter collisions effectively 'short out' part of the magnetospheric electric field. The greater the ionospheric conductivity, the lower the total electric potential the solar wind can maintain across the magnetosphere.   As the electrical conductivity in the polar cap region falls, the potential drop across the magnetosphere approaches the maximum value (which is just the width of the magnetosphere times the -V x B electric field (Slavin et al, 1997).

At Mercury, where no significant ionosphere was identified, currents must close within the regolith, which is a poor conductor at best. The low conductivity of the regolith (about 1 mho based on a lunar sample analysis) makes 'line-tying' negligible and results in short-lived current lines (Encrenaz et al, 2003). As a result, the full, solar wind imposed, cross magnetospheric drop contributes greatly to the short dissipation times predicted by Siscoe et al, 1997 (see above). Regolith-magnetosphere interactions, and the resulting production of hydrogen or hydrogen-bearing compounds and larger volatiles (Gubbins, 1977) are discussed at greater length in Chapter 5 (Potter and Morgan, 1985, Cheng et al, 1987, Ip, 1987, Hunten et al, 1988 and Killen and Morgan, 1993).

6.11   FIELD ALIGNED CURRENTS

In a comparison between the terrestrial magnetosphere and that of Mercury, Hill et al, 1976 suggested that, because of the assumed weaker conductivity of the surface (and the near surface layer) of Mercury, any Field Aligned Current flows should be considerably weaker than those measured in the Earth's magnetosphere. Hill et al (1976) asserted that the solar wind induces the rapid convection leading to substorms in the tail and particle acceleration up to energies of 13 keV. Alternatively, Slavin et al (1997) have reported surprisingly strong field-aligned currents in the near tail following a substorm injection, with total currents into the ion-exosphere similar to such currents at the Earth.

Three principal Field Aligned Current (FAC) systems are commonly observed at the Earth (Slavin, 2004). Region 1 currents are driven by the interaction of the solar wind with the geomagnetic field at the magnetopause and its boundary layers. Region 2 currents are driven by plasma pressure gradients in the inner magnetosphere. Region 3 currents, which are commonly called Substorm Current Wedge (SCW) currents, occur in the midnight local time sector and are generated by the pressure gradients generated in the near-tail by the braking of Earthward directed high speed flows out of the Near Earth Neutral Line (NENL). Both the Region 1 and SCW currents are directed downward towards the planet on the dawnside and outward on the duskside.

Figure 6-17 (Slavin, 2004) depicts region 1 and SCW currents at Mercury. The small scale of Mercury's magnetosphere precludes the occurrence of Region 2 currents. The mechanism by which the Region 1 currents close is presently not understood. Also represented are solar wind induction currents that are driven in the planetary interior by solar wind pressure variations (Glassmeier, 2000 and Suess and Goldstein, 1979). These induction currents act to oppose rapid changes in the pre-existing magnetic field configuration.

A FAC signature can be seen in **Figure 6-19** (Slavin, 2004), a high time resolution segment of the magnetic field measurements taken during Encounter 1 at 1.84 R_M. Following the depolarization event, two minutes after the energetic particle injection event shown in **Figure 6-6**, the By component of the magnetic field went through a very pronounced bipolar variation, the largest variation in this component of the field recorded during Encounters I and III. This large By signature was accompanied only by small variations in the Bx and Bz components and the background field was mostly in the. X_{ME}-Z_{ME} plane. The average Bz around this event was about twice the magnitude of Bx, as would be expected in the case of a pass

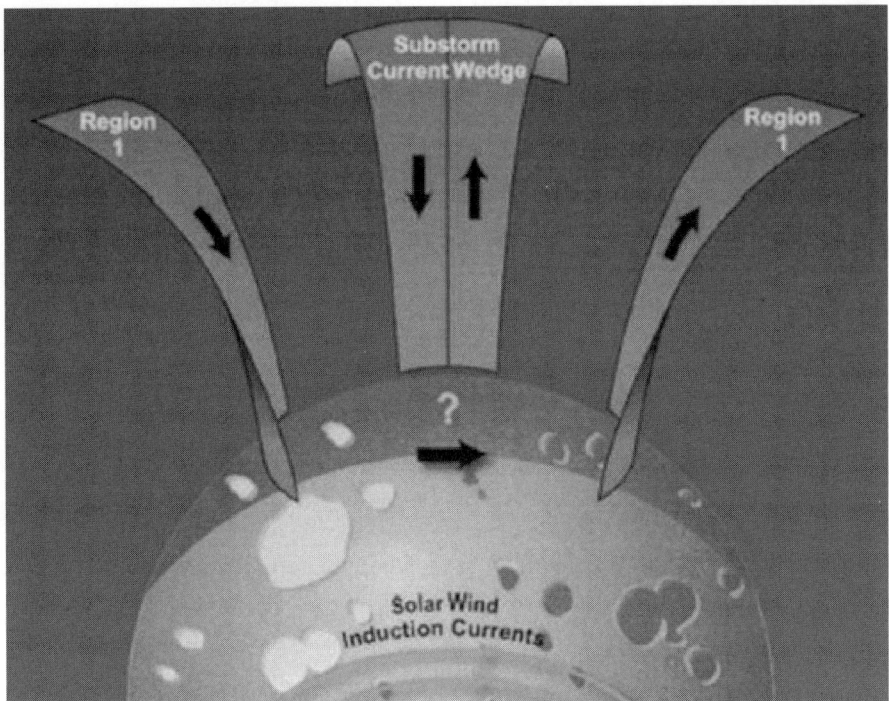

Figure 6-17. Schematic view of Substorm Electrodynamic Interaction between Mercury and its Magnetosphere showing Substorm Current Wedge (SCW) currents in green and solar wind induction currents in orange (Reprinted from Slavin, 2004, with permission from Elsevier).

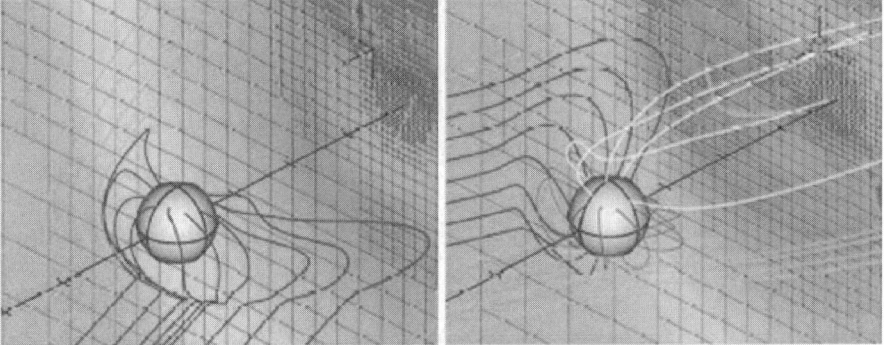

Figure 6-18. MHD Simulation of Mercury's Magnetosphere (Reprinted from Kabin et al, 2000, with permission from Elsevier). Above is view of northern hemisphere, below view of southern hemisphere. The equator and the sub-solar and terminator meridians are indicated on the surface of the planet. The background color reveals the logarithm of magnetic field intensity in a cross-tail plane, 8RM downstream. Magnetic field lines are color coded: tan for IMF lines, magenta for closed field lines, blue and white for open field lines originating in the north and south hemispheres respectively). The simulation reveals a flux tube topology.

through the terrestrial mid-tail region at a distance approaching geosynchronous orbit.

This event is attributed (Slavin, 2004) to the crossing by the spacecraft of a central, quasi-planar, current sheet in which the current flow was nearly aligned with the ambient magnetic field. Given the Mariner 10 trajectory, the main gradient in By from negative to positive was deemed to be indicative of an upward FAC at the spacecraft. This upward current sheet was largely balanced by two smaller downward current sheets before and after the central current sheet, analogous to the occurrence of multiple current sheets at the Earth on the night side of the auroral oval during substorms (Iijima and Potemra, 1978).

Given the strong central current sheet directed into the planet's northern auroral zone at a point well to the east of midnight, the FACs identified might be associated with the Region 1 currents that flow into the poleward edge of the auroral zone in the dawn hemisphere (Slavin et al, 1997). Alternatively, they might be associated with the eastern leg of the Substorm Current Wedge that produced the magnetic depolarization event that occurred earlier (McPheron et al, 1973), followed by an energetic particle injection.

If the observed magnetic field perturbation is assumed to be caused by a semi-infinite current sheet, the implied current density is ~50 mA/m. If this current sheet was quasi-aligned with a constant L shell, then the average current intensity in the central current sheet would be approximately 0.7 $\mu A/m^2$ (Slavin et al, 1997). This calculation assumes that the current sheet was stationary during the very oblique transit by Mariner 10. It is, however, more likely that some dynamic event caused the current sheet to move and precipitate a rapid crossing at the observed time. Nevertheless, both the sheet current intensity and current density inferred are within the range of values observed in the terrestrial magnetosphere at ionospheric to near tail altitudes (Iijima and Potemra, 1978, Elphic et al, 1987).

Assuming that the total length of the current system can be described as a circle with a radius of about 2 R_M, Slavin et al (1997) estimated that the total current, J, flowing into the auroral oval to be 1.4 x 10^6 A during this event. When compared with the total field aligned current flows in the Earth's ionosphere, (~1.2 x 10^6 A), this value appears to be very large given that no conducting ionosphere provides closure (Glassmeier, 2000). It might be argued that, since this is a localized value the total magnitude might be far smaller. Nevertheless, this observation suggests that the electrodynamic coupling between the planet and its magnetosphere is efficient and that conductivity could be high at low altitudes.

One way out of the dilemma is to look for an alternative mechanism that could provide a pathway for the current flow perpendicular to the surface magnetic field. The pick-up current discussed in relation to the Galilean

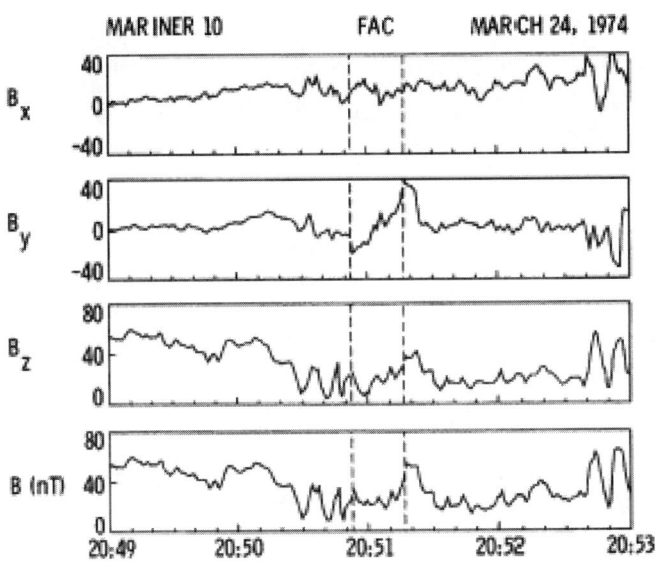

Figure 6-19. Magnetic Field Measurements of Field Aligned Current Event (FAC) observed aboard Mariner 10 during Encounter I (Reprinted from Slavin, 2004, with permission from Elsevier).

satellite Io is a potential candidate that has already been considered applicable to Mercury (Ip and Axford, 1980, Goertz, 1980, and Cheng et al, 1987).

The process could be described as follows. For each (sodium) atom ionized, the initial displacement of the newly created ion and electron represent an electric current. The consequence is that a cross field current flow will be supported by the pickup mechanism. The equivalent, integrated Pedersen conductivity can be estimated (Goertz, 1980) to be $\Sigma_p = N_o\, m_i c^2$ $/\tau_i B^2$ (where N_o is the column density of the neutral sodium atoms, m_i is the atomic mass, c is the speed of light, τ_i is the ionization time scale and B is the surface magnetic field). On substituting for these quantities a value for $\Sigma_p \sim 0.1\text{-}0.3$ mho is obtained. This integrated Pedersen conductivity may be compared with the corresponding value characterizing the dayside terrestrial ionosphere (1 – 10 mho). Ip and Kopp (2004) noted that the patchy nature of particle sputtering with corresponding patchiness in the photo-ionization of pickup ions at Mercury's surface, could result in a terrestrial-like value for Σ_p.

In a preliminary study, Ip and Kopp (2004) adopted a resistivity value appropriate to a fully conducting ionosphere and carried out a set of resistive MHD calculations to see if the field aligned current system might be significantly affected by the orientation of the interplanetary magnetic field.

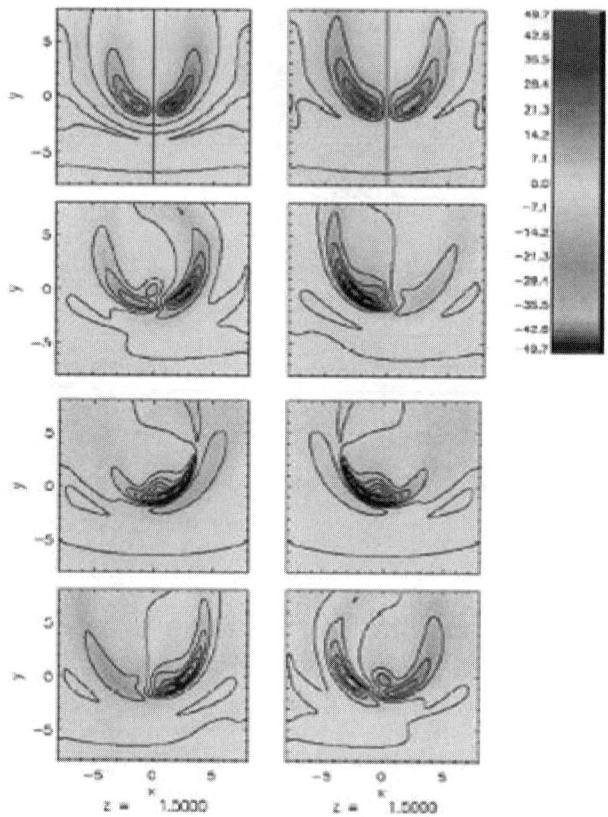

Figure 6-20. Isodensity Profiles of the FAC Events on a plane at a perpendicular distance of 1.5 Rm from the equatorial plane. The solar wind is in the Y direction, the magnetic dipole parallel to the rotation axis in the Z-direction. Current density is in mA/m2. Diagrams are ordered according to the rotational angle of the IMF in the X-Z plane: on left (northward pointing), top to bottom, θ = 0, 45, 90, and 135 degrees; on right (southward pointing), top to bottom, θ = 180, 225, 270, and 315 degrees.   (Reprinted from Ip and Kopp, 2004, with permission from Elsevier.)

This relationship is illustrated by isodensity profiles of the field aligned current (**Figure 6-20**) at a perpendicular distance of 1.5 R_M from the equatorial plane, where the magnetic dipole is assumed to be parallel to the rotational axis in the z-direction. The diagrams are ordered according to the rotational angle of the interplanetary magnetic field in the x-z plane.

An interesting result is that, by assuming a partly conducting ionosphere, it was possible to deduce from the MHD equations the total magnitude of the FAC at the polar region which turned out to be of the order of 10^5 A. This is a factor of ten smaller than the terrestrial value and is correspondingly smaller than the value that emerged from the analysis of Mariner 10 data by Slavin et al (1997). Ip and Kopp (2004) pointed out that even with such a reduced value for the FAC, plasma instabilities might take place, leading to the creation of an electrostatic double layer along the magnetic field lines mapped to Mercury's surface. The resulting particle acceleration could, in turn, lead to enhanced surface sputtering effects in the footprints of the electrostatic double layers.

Given the short time scales for magnetic flux cycling and the absence of terrestrial-style 'line-tying' already noted above, Luhmann et al (1998) suggested that substorms at Mercury may be largely 'driven' by the solar wind. While it is the case that the terrestrial magnetosphere can store magnetic flux for variable amounts of time in the tail lobes before the system becomes unstable and Near Earth Neutral Line (NENL) formation takes place, there are occasions when the magnetosphere dissipates the energy drawn from the solar wind in a fashion that follows the inferred dayside reconnection rate in a linear manner with a fixed phase lag (Bargatze et al, 1985). These latter substorms are classified as 'driven'. The possibility exists that Mercury may not be able to store magnetic flux in its tail due to the absence of an electrically conducting ionosphere. This effect would preclude the occurrence at Mercury of the growth phase often seen to be present in terrestrial substorms (as illustrated in **Figure 6-14**). However, as argued by Slavin (2004) in conformity with earlier work (Slavin and Holzer, 1979), there would still be some minimal convection time delays in the magnetosheath and then the tail (of the order of a few minutes) as magnetic flux is transported to and from the dayside magnetosphere.

Goldstein et al (1981) calculated the distribution of the solar wind stand-off distance at Mercury as a function of solar wind dynamic pressure for two reconnection, or flux-transfer, scenarios (**Figure 6-21**): Earth-like flux transfer, and no flux transfer. The stand-off distances are obviously affected by the nature of reconnection and have a narrower distribution and higher minima for the no flux transfer case.

Potential solar wind plasma access and acceleration processes taking place in Mercury's magnetosphere are illustrated to a first approximation in **Figure 6-22** (Lundin et al, 1997). Acceleration processes include current sheet, field aligned and transverse processes analogous to acceleration/heating processes in the top of Earth's ionosphere.

6.12 DETECTABLE MAGNETOSPHERE/EXOSPHERE INTERACTIONS

What is the role of the exosphere as a source of plasma for the magnetosphere? Ionization of the exosphere will create detectable amounts of low energy plasma. Solar wind-induced electric fields either accelerate ions into the magnetosphere or back to the surface where they are

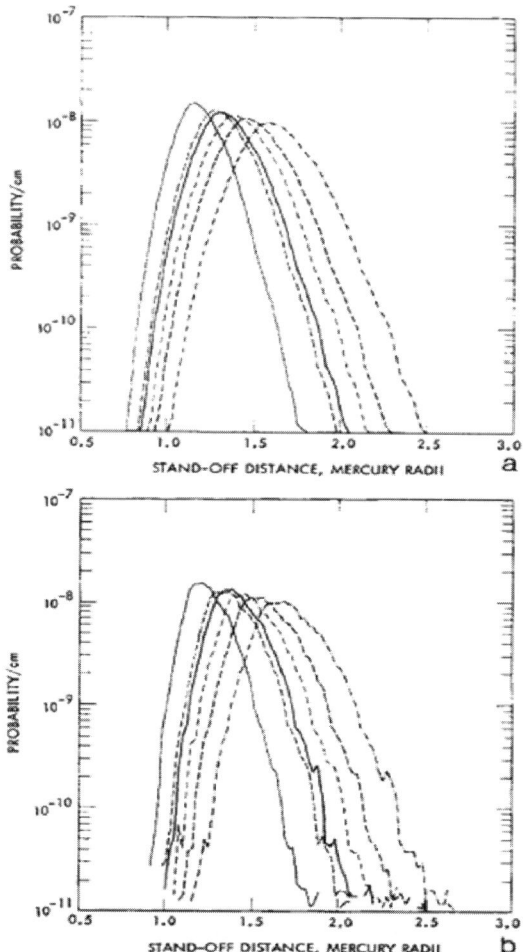

Figure 6-21. Distribution of Solar Wind Standoff Distances as a function of solar wind dynamic pressure for (a) Earth-like flux transfer and (b) no flux transfer. Heavy lines are aphelion cases, thin lines perihelion cases. Solid, dashed, and dash-dot curves correspond respectively to magnetic moments M = 2.5, 3.5 and 4.5 x 1022 g cm-3. (Goldstein at el, 1981, copyright AGU. Used with permission of AGU.)

reimplanted. In the extreme case, when solar wind impinges directly on the surface, the ions would gain energy at a rate of a few eV per kilometer, resulting in energies in the keV range near the surface. These topics are discussed in more detail in Chapter 5.

In any case, energetic neutral atoms produced as a result of ion exchange between proton plasma in the magnetospheric energy range and the surrounding neutral gas are detectable and thus provide a measure of such interactions. Neutral gas produced by a variety of mechanisms discussed in Chapter 5 includes, Hydrogen, Helium, and hydrogen-bearing compounds.

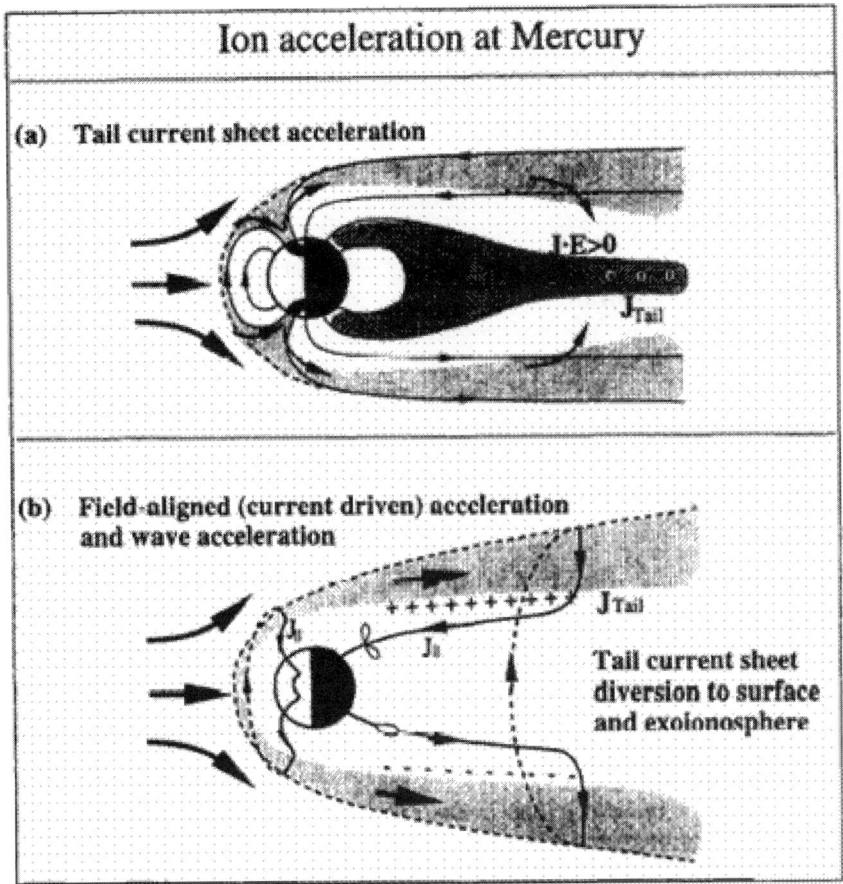

Figure 6-22. Solar Wind Interactions with Mercury's Magnetosphere, involving plasma access and acceleration processes by a) current sheet, and b) field-aligned and transverse processes. Processes are analogous to acceleration/heating processes in the topside ionosphere of the Earth. (Reprinted from Lundin et al, 1997, with permission from Elsevier.)

Ions, of these and other elements, including O, Na, and K, are created on the dayside by a combination of mechanisms including UV and solar wind ionization processes, and then become trapped in the magnetosphere (Kallenrode, 2004). These ions can be detected and used to 'image' the proton plasma injected and trapped in the magnetosphere (Brandt et al, 1999).

Mercury's relatively small magnetosphere, slow rotation, and thin atmosphere prevent the generation of a corotational electric field and stably trapped particles or plasmasphere (Russell et al, 1988). Effectively, the corotation region has a smaller radius than the planet so that closed field lines exist only near the equator (Encrenaz et al, 2003). This makes for a very dynamic system dominated by movement and reconnection on open field lines. A strong solar wind-magnetosphere coupling, the escape of volatiles along open field lines, and very brief duration events associated with solar wind induced magnetic convection would be anticipated characteristics (Encrenaz et al, 2003). Sufficiently energetic protons can be injected and quasi-trapped close to the nightside surface. During magnetic storms on Earth, such plasma would drift inward, create an intensified ring current, and relax due to charge exchange with the ionosphere. On Mercury, where substorms are far more frequent (10/hour) and shorter in duration (tens of seconds), storm-injected plasma relaxes rapidly through direct interaction with the surface, creating a 'flash' as charge exchange occurs in a matter of seconds (Brandt et al, 1999). The spectra and timescale of ion exchange is characteristic, and thus could be diagnostic, for each atmospheric species (Brandt et al, 1999).

In order to simulate the expected energetic neutral atom fluxes at Mercury, Lukyanov et al (1997) developed a model of the energetic ion dynamics by adapting the magnetic field model of Tsyganenko (1989) scaling it in absolute dimensions so that 1 R_M = 8 R_E. Only a dawn-dusk electric field set up by the solar wind was assumed, (i.e., no co-rotational field was included due to the slow rotation of the planet). A solar magnetic (SM) co-ordinate system was employed (Russell, 1971) with the x-axis pointing towards the Sun and the z-axis aligned parallel to the north dipole axis. No tilt of the dipole field with respect to the ecliptic plane was assumed. The electric and magnetic fields in the model were taken to be static. This latter assumption is valid if the typical travel time of an ion through the magnetosphere is shorter than the typical time scales of field variations. Further it was further assumed that the magnetic field did not change over the gyroradius of an ion (i.e. the guiding center drift approximation was used). If the magnetospheric plasma is assumed to be collisionless, then by Liouville's Theorem the distribution function is invariant along the particle trajectory. An initial distribution function of particles is injected into the magnetosphere and traced through the electric and magnetic fields using the drift trajectory equations. Particles are lost when they reach the boundary of the magnetosphere and when they impact the planet. By maintaining a steady injection rate, a steady state model of the ion distribution can be achieved. **Figure 6-23** is a map of the column integrated ion density with the subsolar point in the center (meridional view). The drift approximation breaks down close to the horizontal center line in the tail and, wherever such a breakdown occurs particles are taken out of the model and considered to be lost. The corresponding simulated ENA images are generated (**Figure 6-23**) by assuming a spinning imager similar to the PIPPI instrument flown on Sweden's ASTRID satellite (Brandt et al, 1999).

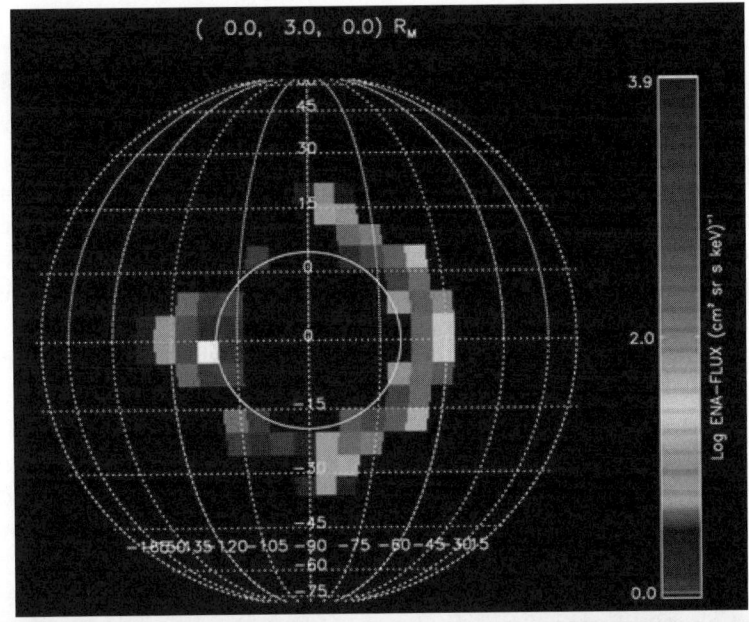

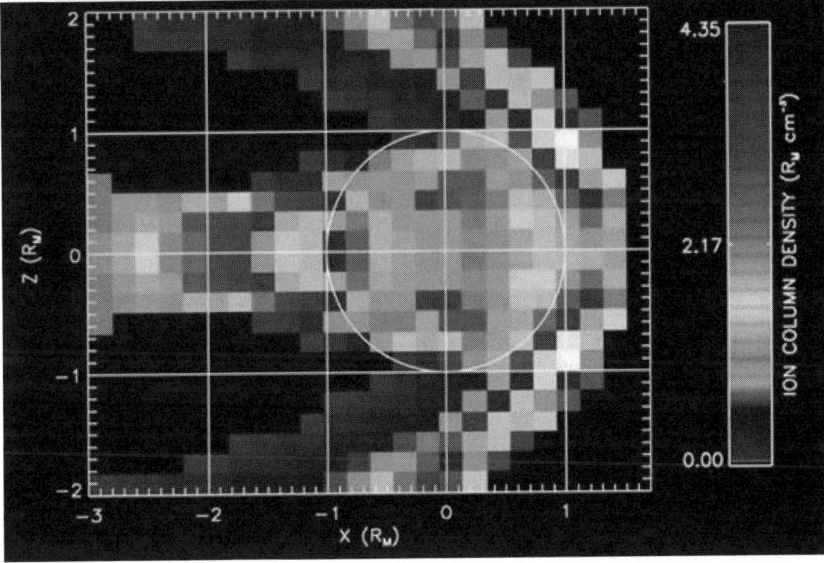

Figure 6-23. Ion Density and Energetic Neutral Atom Distribution. Above, a meridional view of ion column density for 50 keV ions at Mercury integrated along the y axis. Below, Energetic Neutral Atom image obtained with virtual imager with 5 x 5 resolution corresponding to ion model above. Both courtesy of Pontus Brandt. Here the initial energy is 50 keV, the initial differential ion flux 10^6 (cm^2/.keV); the initial ion density 2 cm^{-3} and the electric field was 10 mV/m. All components parallel to the magnetic field are set to zero to avoid electrostatic acceleration. The source region was placed at XSM = 3R**M**, centered on, and normal to, the x axis. The source surface area was 1RM x 1RM

6.13   MAGNETOSPHERE/SURFACE INTERACTIONS

The small scale of Mercury's magnetosphere combined with the presence of a tenuous atmosphere result in the inner magnetosphere lying below the surface (Brandt, 1999). Consequently, 1) a relatively large surface area interacts with plasma-sheet and tail region accelerated particles; and 2) energetic particles have closed drift paths directly into the surface. In addition, the response of the magnetosphere to the highly variable particle pressure leads to modifications in the size and locations of the surface regions exposed to intensive sputtering fluxes. Although the surface is made up of relatively porous regolith and thus not highly conducting (Carrier et al, 1991), it is called upon to play a role analogous to Earth's ionosphere. How is the surface unlike the ionosphere? Grard et al (1997) combined results from experiments and modeling to determine the nature of conductivity in the proximity of Mercury's surface. They concluded that charge exchanges in response to photoemission would occur in the sunlit portion of the surface, supply negative charges only, and result in the flow of electrons. A conductive photosheath layer would be generated above the surface as a result of photoemission. This sheath would provide electron transport more efficient than the height-integrated exosphere alone and as efficient as the regolith integrated over tens of meters and thus would play a primary role in the conductance of electrons. Still, the transport of electrons is several orders of magnitude more efficient in the Earth's ionosphere than at or near Mercury's surface. **Figure 6-24** (Grard, 1997) illustrates the nature of possible current exchanges between the surface and the charged particle environment. Surface emitted electrons may return to the surface. Many escape to maintain the net flow of electrons. Returning electrons can flow as current in closed surface/magnetosphere loops.

6.14   RECENT MODELING OF THE MAGNETOSPHERE

Although the majority of earlier Mercury magnetic field models were 'closed field' models, Hill et al (1976) and Luhmann et al (1998) proposed that reconnection should occur, and that open field lines exist connecting the planet to the solar wind. Because the planet surface is large compared to the size of the dayside magnetosphere, if reconnection does occur, the region of direct solar wind impact on the surface may be exceedingly large and dynamic. Because Mercury is close to the Sun, the local IMF has a dominant radial component resulting in large solar wind fluxes on the northern polar cap for an outward-directed field and a larger solar wind flux on the southern polar cap for an inward-directed field. This effect is illustrated in **Figure 6-**

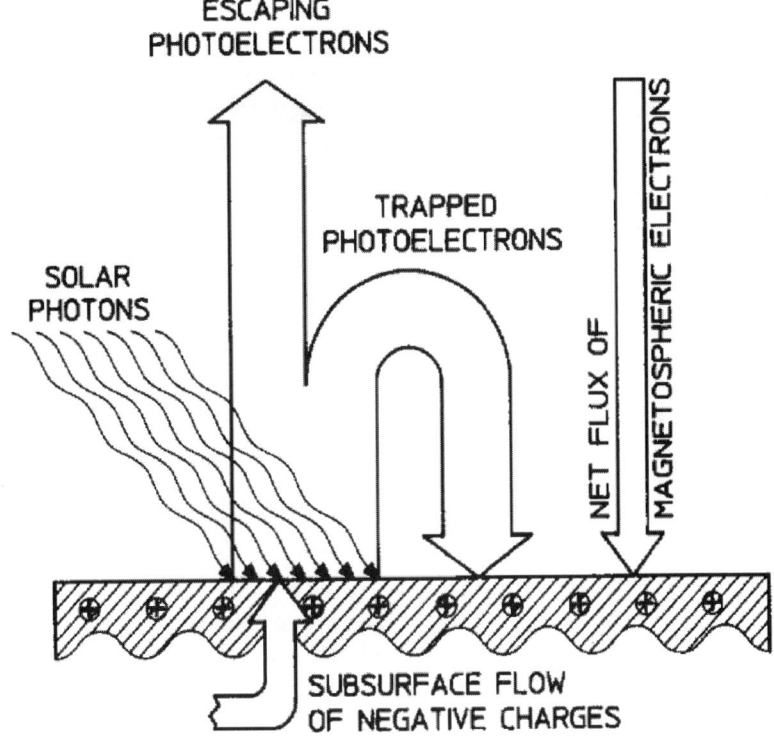

Figure 6-24. Schematic Magnetosphere/Surface Charge Exchange for Mercury. (Reprinted from Grard, 1997, with permission from Elsevier.)

25 (Giles, 1998), where the Toffoletto-Hill magnetic field model is used to predict the field connection and polar cap size for radially outward IMF's. The solar wind flux impacts the entire northern polar cap above about 60 degrees. The caps will have asymmetric footprints with an IMF y-component resulting in an additional dawn-dusk displacement. A southward-directed field, which should be uncommon, yields large, full polar caps in both hemispheres.

Sarantos et al (2001) scaled a well tested model of the terrestrial magnetosphere to Mercury and used it to examine the access of solar wind ions to the surface of the planet, along magnetically open field lines. Based on the findings of Slavin and Holzer (1997) of higher Alfven speeds in the solar wind at 0.3AU, they increased the efficiency of the reconnection process by 40%. The Sarantos et al (2001) model is shown in **Figure 6-26**. This figure shows, projected on the dayside hemisphere, the footprints of recently opened field lines that connect the planetary surface to the solar wind under representative solar circumstances (Bx = - 40 nT, By = - 5 nT, Bz = - 20 nT). . The 'empty' region at low latitudes corresponds to the rather

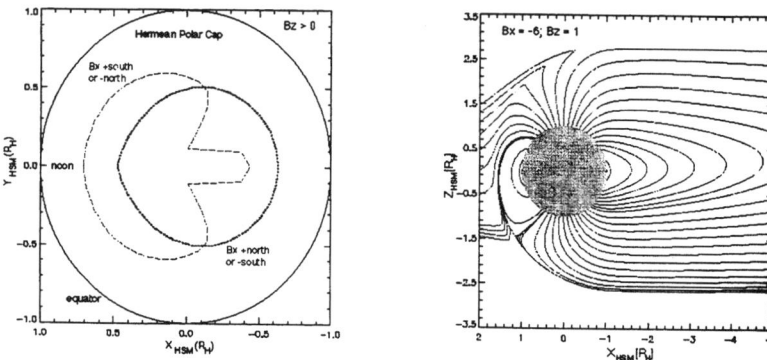

Figure 6-25. Toffoletto-Hill Magnetosphere Model. Predications of Mercury's polar cap (top) and magnetic field connection lines (bottom) for outward (left) and inward (right) directed IMF. Courtesy of Barbara Giles.

small part of the surface shielded from the solar wind by closed magnetic field lines. The model exhibits strong asymmetry. The field lines emerging from the northern/southern hemispheres map to the upstream/downstream solar wind for (significant) negative/positive values of the IMF B_x, component. Only the northern hemisphere (black dots) connects to the upstream solar wind (for strongly negative B_x) and channels solar wind protons to the surface where sputtering may produce an enhanced outward flux of neutral atoms (as discussed in Chapter 5). The hemispheric asymmetries predicted by this model, which could be triggered in practice by the crossing of interplanetary sector boundaries, could contribute to the neutral atom variability reported by Potter et al (1999) and others.

A similar analysis was carried out by Leblanc et al (2003) for ions with energies >10 keV/amu accelerated in association with Coronal Mass Ejections (CMEs). This study also showed that charged particle access to the surface of Mercury depended critically on the IMF and that the fluxes of sputtered materials resulting from related surface impacts would also be highly asymmetric.

Similar features emerged in global MHD and hybrid simulations of the solar wind interaction with Mercury (e.g. Kallio and Janhunen, 2003, Ip and Kopp, 2002 Kabin et al, 2000). MHD model simulations for the northern and southern hemispheres are shown in **Figure 6-18** (Kabin et al, 2000). The model reveals a flux tube topology. In the southern hemisphere, IMF and open field lines originating in the northern hemisphere can be seen in the equatorial plane upstream. Closed field lines originating in the southern polar cap can clearly be seen. Magnetic field lines emanating from the polar regions map into the lobes of the tail while showing a strong asymmetry in the draping of the newly opened flux tubes. In the northern hemisphere, the open field lines originating in the northern hemisphere are uniformly

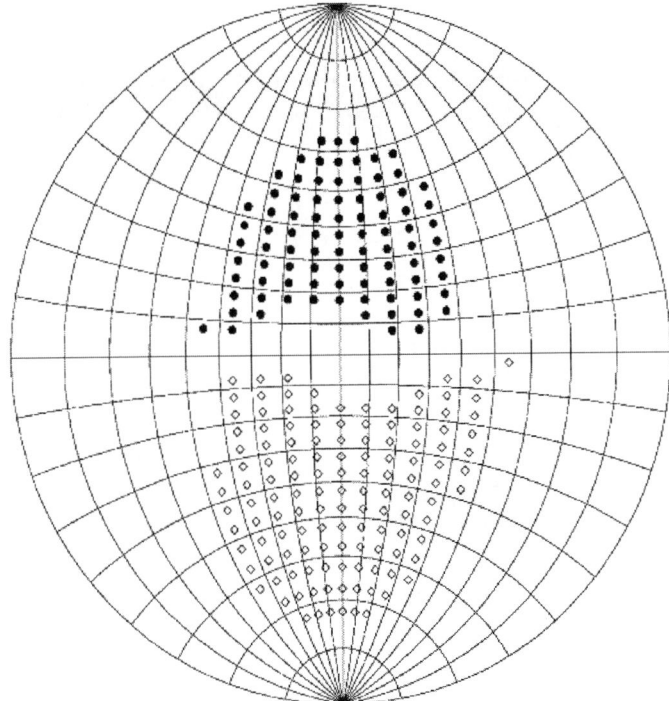

Figure 6-26. Interplanetary Magnetic Field for Hybrid Magnetosphere Model. Regions of open magnetic field mapping to upstream (black dots) and downstream (open diamonds) (Reprinted from Sarantos et al, 2001, with permission of Elsevier).

distributed along the cusp region, and they belong to the different upstream horizontal planes. Closed field lines and open field lines originating in the southern hemisphere are noticeably absent. In both hemispheres, the recently opened open field lines originating in the northern hemisphere connect to the upstream solar wind thus allow the direct impact of solar wind ions on the surface. The open field lines originating in the southern hemisphere map only to the downstream solar wind.

Kallio and Janhunen (2004) developed a 3D quasi-neutral hybrid model of the magnetosphere. **Figure 6-27** illustrates the response of their model to the IMF under three conditions: 1) pure northward directed IMF, 2) pure southward directed IMF, and 3) Parker spiral IMF. The magnetosphere behaves very similarly in the first two cases, with a dawn-dusk asymmetry, but is slightly more open in the pure southward case due to the erosion of the dayside magnetopause. This is consistent with the analyses of Mariner 10 data by Slavin and Holzer (1979) and observations of the terrestrial magnetosphere (Yang et al, 2002). The open field lines extend to lower latitudes and occur over a larger region for the southward directed IMF. These characteristics are consistent with other models (Sarantos et al, 2001); Ip and Kopp, 2002) discussed here. The Parker Spiral case has a pronounced

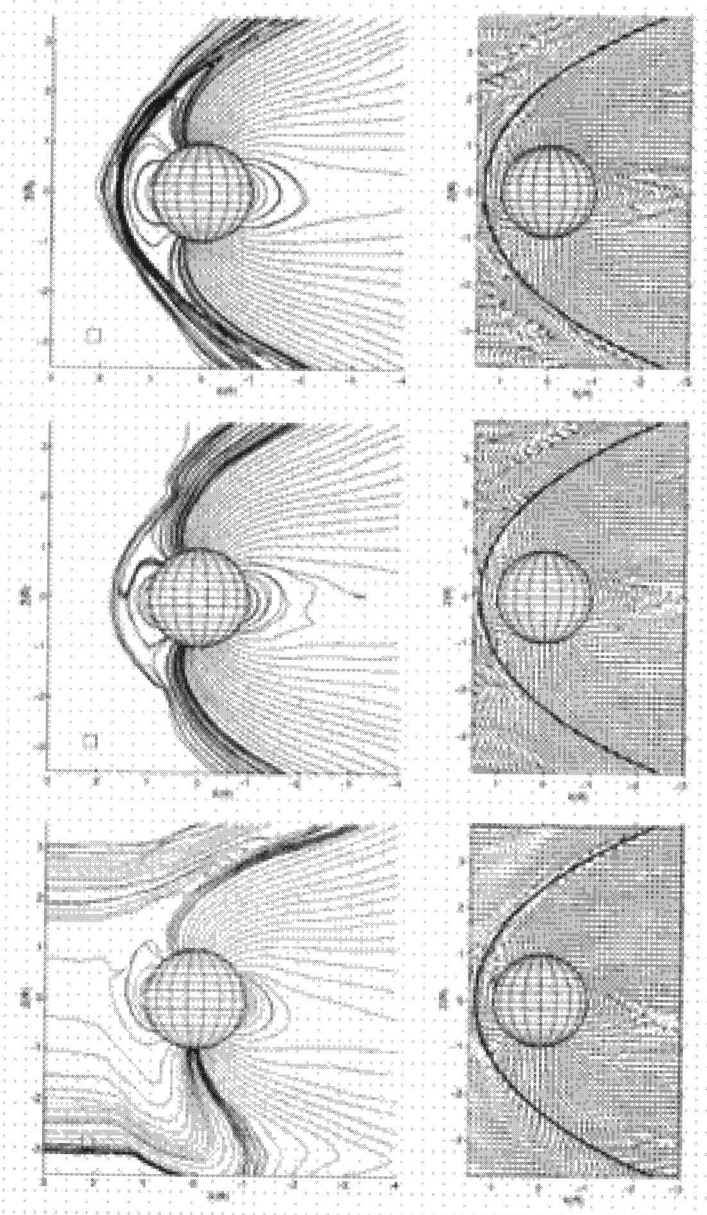

Figure 6-27. Neutral Hybrid Magnetosphere Models. Magnetic field lines (left) and corresponding normalized field vectors (right) at IMF [0,0,10] nT (top), [0,0,-10] nT (middle), and [32,10,0] nT (bottom). See text for details. (Reprinted from Kallio and Janhunen, 2004, with permission from Elsevier.)

north-south asymmetry with one hemisphere strongly connected to the solar wind. This case is similar to the closed field model developed to fit the Mariner 10 data (Whang, 1977).

Mercury's magnetosphere has characteristics in the 'intermediate' range (Kallio and Janhunen, 2004) between pure IMF controlled induced magnetospheres at one extreme, as exemplified at Venus and Mars, and intrinsic magnetic field controlled magnetosphere at the other extreme, as exemplified at the Earth and other planets.

6.15   SUMMARY

Mercury is in a unique position to provide understanding of planetary magnetospheres. The planet has a simple magnetosphere formed by the solar wind, which shapes it; the interplanetary magnetic field (IMF), which activates it; an interior magnetic dipole, which structures it; and the planet itself which anchors it. These are the basic ingredients. All other known magnetosphere-bearing planets have added complexities, such as additional sources of plasmas or current systems, that encumber the understanding of magnetospheric processes. Mercury's relative inaccessibility has prevented us from fully documenting its basic 'load free' magnetosphere. However, modeling efforts, based on the development of codes for the Earth's magnetosphere, have attained a high level of maturity, sophistication, and power, and have been used extensively to enhance our understanding of the relationship between measurements and their implications.

6.16   REFERENCES

Angelopoulos, V.., w. Baumjohann, C. Kennel, et al, Bursty bulk flows in the inner central plasma sheet, *JGR*, **97**, 4027-4039, 1992.

Armstrong, T., S. Krimigis, L. Lanzerotti, A reinterpretation of the reported energetic particle fluxes in the vicinity of Mercury, *JGR*, **80**, 4015-4017, 1975.

Anderson, J. D., G. Colombo, P. B. Esposito, E. L. Lau, and G. B. Trager, The mass, gravity field, and ephemeris of Mercury, *Icarus,* **71**, 337-349, 1987.

Baker, D., T. Pulkkinen, V. Angelopoulos, W. Baumjohann, R. McPherron, Neutral line model of substorms: past results and present view, *JGR*, **101**, 12975-13010, 1996.

Baker, D., J. Simpson, J. Eraker, A model of impulsive acceleration and transport of energetic particles in Mercury's magnetosphere, *JGR*, **91**, 8742-8748, 1986.

Baker, D., J. Borovasky, J. Burns, G. Gisler, M. Zellie, Possible calorimetric effects at Mercury due to solar wind magnetospheric interactions, *JGR Space Physics*, **92**, 4707-4712, 1986.

Barabash, S., A. Lukianov, P. C. Brandt and R. Lundin, Energetic neutral atom imaging at Mercury: Simulations and instrument requirements, *IRF Scientific Report* **244**, 51-64, 1997.

Bargatze, L., D. Baker, R. McPherron, E. Hones, Magnetospheric impulse response for many levels of geomagnetic activity, *JGR*, **90**, 6387-6394. 1985.

Blomberg, L. and J. Cumnock, On electromagnetic phenomena in Mercury's magnetosphere, *Adv. Space Res.,* **33**, 2161-2165, 2004.

Brandt, P., ENA Imaging of Planetary Magnetospheres, Doctoral Thesis at the Swedish Institute of Space Physics, Kiruna, 1999.

Brandt, P., S. Barabash, O. Norberg, R. Lundin, E.C. Roelof and C.J. Chase, Energetic neutral atom imaging at low altitudes from the Swedish microsatellite Astrid: images and spectral analysis, *JGR*, **104**, 2367-2379, 1999.

Campbell, W.H., Introduction to Geomagnetic Fields (Publ. Cambridge U. Press), 291 p., 1997.

Cheng, A. R. Johnson, S. Krimigis, L. Lanzerotti, Mercury's magnetosphere, exosphere, and surface low frequency field and wave measurements as a diagnostic tool, *Planet. Space Sci.,* **45**, 143-148, 1987.

Christon, S., A comparison of the Mercury and Earth magnetospheres: Electron measurements substorm time scales, *Icarus*, **71**, 448-471, 1987.

Connerney, J. E. P., and N. F. Ness, Mercury's Magnetic Field and Interior, in *Mercury*, F. Vilas, C. Chapman, and M. Matthews, Eds., Univ. Arizona Press, pp. 494-513, 1988.

Connerney, J. E. P., M.H. Acuna., P.J. Wasilewski, G. Kletetschka, N.F. Ness, H. Reme, R.P. Lin and D.L. Mitchell, The Global Magnetic Field of Mars and Implications for Crustal Evolution, *GRL,* **28**, Issue 21, 4015-4018, 2001.

Elphic, R., D. Southwood, Simultaneous measurements of the magnetopause and flux-transfer events at widely separated sites by AMPTE UKS and ISEE-1 and ISEE-2, *JGR*, **92** (A12), 13666-13672, 1987.

Encrenaz, T., J.P. Bibring, M. Blanc, M.A. Barucci, F. Roques, P. Zarka, *The Solar System*, (Publ. Springer), 512 p., 2003.

Eraker, J. and J. Simpson, Acceleration of charged particles in Mercury's magnetosphere, *JGR Space Physics*, **91**, 9973-9993, 1986.

Glassmeier, K. H., Concerning substorms in the Hermean magnetosphere, Proc. 3rd. International Conference on Substorms (ICS-3) Versailles, *ESA SP*-**389**, 707-712, 1996.

Glassmeier, K. H., The Hermean magnetosphere and its ionosphere-magnetosphere coupling, *Planet. Space Sci.,* **45**, 119-125,1997.

Glassmeier, K. H., Current Systems in Mercury's Magnetosphere, Proc. of the Chapman Conference on Magnetospheric Currents, AGU Washington, 2000.

Goertz, C.K., Io's Interaction with the Plasma Torus, *JGR-Space Physics*, **85**, NA6, 2949-2956, 1980.

Goldstein, B., S. Suess, R. Walker, Mercury: Magnetospheric processes and atmospheric supply and loss rates, *JGR Space Physics*, **86**, 5485-5499, 1981.

Grard, R., Photoemission on the surface of Mercury and related electrical phenomena, *Plan Space Sci*, **45**, 1, 67-72, 1997.

Gubbins, D., Speculations on the Origin of the Magnetic Field of Mercury, *Icarus*, **30**, 186-191, 1977.

Hamilton, D., F. Gloeckler, M. Ipavich, W. Studemann, B. Wilkin, G. Kremser, Ring current development during the great geomagnetic storm of February 86, *JGR,* **93**, 14343-14355, 1988.

Hood, L. and G. Schubert, Inhibition of solar win impingement on Mercury by planetary induction currents, JGR, 84, 2641-2647, 1997.

Hunten, D. M., T. H. Morgan, and D. E. Shemansky, The Mercury atmosphere, in *Mercury,* University of Arizona Press, Tucson., 562-613, 1988.

Hill, T. W., A. J. Dessler, and R. A. Wolf, Mercury and Mars: The role of ionospheric conductivity in the acceleration of magnetospheric particles, *JGR*, **3**, 429-432, 1976.

Iijima, T. and T. Poterma, Large scale characteristics of field-aligned currents associated with substorms, *JGR*, **83**, 599-615, 1978.

Ip, W., Dynamics of electrons and heavy ions in Mercury's magnetosphere, *Icarus*, **71**, 441-447, 1987.

Ip, W. and W. Axford, Weak interaction-model for Io and the Jovian magnetosphere, *Nature*, **283**, 5743, 180-183, 1980.

Ip, W. and A. Kopp, Mercury's Birkeland current system, *Adv. Space Res.,* **33**, 2172-2175, 2004.

Kabin, K. and T. Gombosi et al, Interaction of Mercury with the solar wind, *Icarus,* **143,** 397-406, 2000.

Kallenrode, M.B., *Space Physics* (Publ. Springer), 482 p., 2004.

Kallio, E. and P. Janhunen, Modeling the solar wind interaction with Mercury by a quasineutral hybrid model, *Ann Geophys*, **21** (11), 2133-2145, 2003.

Kallio, E. and P. Janhunen, The response of the Hermean magnetosphere to the interplanetary magnetic field, *Adv in Space Research*, **33**, 2176-2181, 2004.

Killen, R. and T. Morgan, Maintaining the Na atmosphere of Mercury, *Icarus,* **101,** 293-312, 1993.

Killen, R. and T. Morgan, Diffusion of Na and K in the uppermost regolith of Mercury, *JGR,* **98**, 23589-23601, 1993.

LeBlanc, F., J. Luhmann, R. Johnson, M. Lui, A Solar energetic particle event at Mercury, *Planet Space Sci,* **51**, 339-352, 2003.

Lewis, Z., S. Cowley, D. Southwood, Impulsive energization of ions in the near-Earth magnetotail during substorms, *Planet Space Sci,* **38**, 491-505, 1990.

Luhmann, J., C. Russell, N. Tsyganenko, Disturbances in Mercury's magnetosphere: are the Mariner 10 substorms simply driven, *JGR,* **103**, 9113-9119, 1998.

Lukyanov,A.V., S. Barabash, R. Lundin and P.C. Brandt, A model of energetic charged particle dynamics in the Mercury magnetosphere, Proc. Mercury Workshop. Ed. S. Barabash, (Publ. Swedish Institute of Space Physics), *IRF Scientific Report* **244**, 35-50, 1997.

Lundin, R., S. Barabash, P. Brandt, L. Eliasson, C. Nairn, O. Norberg, I. Sandahl, Ion acceleration processes in the Hermean and terrestrial magnetospheres, *Adv Space Res,* **19** (10), 1593-1607, 1997.

McPherron, R., C. Russell, M. Aubry, Satellite studies of magnetospheric substorms on August 15, 1968, 9. Phenomenological model for substorms, *JGR,* **78**, 3131-3149, 1973.

Ness, N., K. Behannon, R. Lepping, Y. Whang, and K. Schatten, Magnetic field observations near Mercury: Preliminary results from Mariner 10, *Science,* **183**, 1301-1306, 1974.

Ness, N., K. Behannon, R. Lepping, Y. Whang, and K. Schatten, The magnetic field of Mercury, *JGR,* **80**, 2708, 1975.

Ness, N., K. Behannon, R. Lepping, Y. Whang, Observations of Mercury's magnetic field, *Icarus,* **28**, 479-488, 1976.

Ness, N., The magnetosphere of Mercury, in *Solar System Plasma Physics,* **1**, Eds Kennel, C., L. Lanzerotti, E. Parker, North Holland, Amsterdam, 183-206, 1979.

Ogilvie, K., J. Scudder, R. Hartle, G. Siscoe, H. Bridge, A. Lazarus, J. Asbridge, S. Bame, C. Yeates, Observations at Mercury encounter by the plasma science experiment on Mariner 10, *Science,* **185**, 145-151, 1974.

Ogilvie, K., J. Scudder, V. Vasyliunas, R. Hartle, G. Siscoe, Observations at the planet Mercury by the plasma electron experiment: Mariner 10, *JGR,* **82**, 1807-1824, 1977.

Okuchi, T., Hydrogen partitioning into molten iron at high pressure: Implications for Earth's core, *Science,* **278**, 1781-1785, 1997.

Peale, S. J., Does Mercury have a molten core?, Nature, 262, 1976.

Potter, A. and T. Morgan, Discovery of sodium in the atmosphere of Mercury, *Science,* **229**, 651-653, 1985.

Quinn, J. and C. McIlwain, Bouncing ion clusters in the Earth's magnetosphere, *JGR,* **84**, 7365-7370, 1979.

Quinn, J. and D. Southwood, Observations of parallel ion energization in the equatorial region, *JGR,* **87**, 10536-10540, 1982.

Richardson, I. G., Simultaneous plasma wave, magnetic field, and energetic ion observations in the ion pickup region of comet P/Giacobini-Zinner, Proceedings of Chapman Conference on Plasma Waves and Instabilities in Magnetospheres and at Comets, Sendai, Japan, 1987.

Russell, C.T., D.N. Baker, and J.A. Slavin, The Magnetosphere of Mercury. In *Mercury* (Eds. F. Vilas, C. Chapman and M. Mathews) Univ. of Arizona Press, Tucson, 514-561, 1988.

Russell, C. T., Magnetic fields of the Terrestrial Planets, *JGR,*, **98**(E), 18681-18695, 1993.

Russell, C., ULF waves in the Mercury magnetosphere, *GRL*, **16**, 1253-1256, 1989.

Russell, C., On the relative locations of the bow shocks of the terrestrial planets, *GRL*, **4**, 387-390, 1977.

Russell, C.T., Geophysical co-ordinate transformations, in *Cosmic Electrodynamics,* **2**, (Publ. D. Reidel Publ. Co.), 184-196, 1971.

Russell, C. and R. McPherron, Semiannual variation of geomagnetic activity, JGR, **78**, 92-108, 1973.

Russell, C. and R. Walker, Flux transfer events at Mercury, *JGR*, **90**, 11067-11074, 1985.

Sarantos, M., P. Reiff, T. Hill, R. Killen, A. Urquhart, A Bx interconnected magnetosphere model for Mercury, *Planet Space Sci*, **49**, 1629-1635, 2001.

Schubert, G., M. N. Ross, D. J. Stevenson, and T. Spohn, Mercury's thermal history and the generation of its magnetic field, in *Mercury,* eds. F. Vilas, C. Chapman, and M. Matthews, Univ. Arizona Press, 429-460, 1988.

Siscoe, G. and L. Christopher, Variations in Solar Wind Stand-Off Distances at Mercury, *GRL,* **4**, 1580169, 1975.

Shiakawa, K., W. Baumjohann, G. Haerendel, High speed ion flow, substorm current wedge formation and multiple Pi2 pulsations, *JGR*, **103**, 4491-4507, 1998.

Sigfried, R. W., and S. C. Solomon, Mercury: Internal Structure and thermal evolution, *Icarus,* **23**, 1974.

Simpson, J., J. Eraker, J. Lamport, P. Walpole, Electrons and protons accelerated in Mercury's magnetic field, *Science*, **185**, 160-166, 1974.

Siscoe, G. and I. Christopher, Variations in the solar wind standoff distance at Mercury, *GRL*, **2**, 158-160, 1975.

Siscoe, G., N. Ness, C. Yeates, Substorms on Mercury?, *JGR*, **80**, 4359-4363, 1975.

Siscoe , G., Magnetospheric physics-Big storms make little storms, *Nature,* **390**, 6659, 448-449, 1997.

Slavin, J. and R. Holzer, The effect of erosion on the solar wind standoff distance at Mercury, *JGR,* **84**, 2976-2082, 1979.

Slavin, J., C. Owen, J. Connerney, S. Christon, Mariner 10 observations of field-aligned currents at Mercury, *Planet Space Sci*, **45**, 133-141, 1997.

Slavin, J., D. Fairfield, R. Lepping, Simultaneous observations of earthward flow bursts and plasmoid ejection during magnetospheric substorms, *JGR*, **107** (A7), doi:10.1029/2000JA003501, 2002.

Slavin, J., Mercury's magnetosphere, *Adv in Space Res*, **33**, 1859-1874, 2004.

Solomon, S., Some aspects of core formation in Mercury, *Icarus*, **28**, 509-521, 1976.

Southwood, D. J., The magnetic field of Mercury, *Planet. Space Sci.*, **45**, 113-118, 1997.

Suess. S. and B. Goldstein, Compression of the Hermean magnetosphere by the solar wind, *JGR*, **84**, 3306-3312, 1979.

Tsyganenko, N.A., A magnetospheric magnetic field model with a warped tail current sheet, *Planet. Space Sci.*, **37**, 5, 1989.

Whang, Y., Magnetospheric magnetic field of Mercury, *JGR Space Physics*, **82**, 7, 1024-1030, 1977.

Yang, Y.H., Comparison of three magnetopause prediction models under extreme solar wind conditions, *JGR Space Physics*, **107**, SMP3-1 to SMP3-9, 2002.

6.17   SOME QUESTIONS FOR DISCUSSION

1. How can Mercury expand our understanding of planetary magnetospheres?

2. Compare and contrast the magnetospheres of Mercury, the Earth, and Jupiter.

3. What in situ measurements are essential in understanding the spatial and temporal structure of a magnetosphere?

4. What are the consequences of size of Mercury's field on the structure of its magnetosphere and the interaction of plasma with other subsystems?

Chapter 7

THE FUTURE OF MERCURY EXPLORATION

7.1   NEED FOR FURTHER INVESTIGATION OF MERCURY'S INTERIOR

Recent experience at Mars (Acuna et al, 1998) illustrates the potential complexity and variety of planetary interiors and magnetic fields. Thus every model, even unlikely crystal magnetization, should be under consideration until data considerably improved over Mariner 10 are obtained by another mission to Mercury (Giampiere and Balogh, 2002). Determination of multipole coefficients must provide the constraints necessary to establish the core as a magnetic dipole. Correlation of gravitation and magnetic fields, due to thermo-electric control of the magnetic field by iregularities, or topography, at the core mantle interface, could support Stevenson's hypothesis of a thermo-electric dynamo (Giampieri and Balogh, 2002). Further characterization of the interior necessitates sending a mission to Mercury to measure the rotation characteristics, including moments of inertia, to an accuracy which constraints the size and state of the core. The structure of the global gravity field must be determined to an accuracy that constrains the deep internal structure and the local gravity anomalies to an accuracy that constrains the nearer-surface structures of the crust and mantle. Much better relative accuracy and signal to noise ratio of better than 1000 is required. C_m/C, needs to be determined to better than 0.05, and C/MR^2, the concentration coefficient, to better than 0.003. Additional data are required to describe the core quantitatively and to identify any variation that would elaborate the nature of the dynamo (Connerney and Ness, 1988).

7.2   GROUND-BASED OBSERVATIONS FOR INTERIOR EXPLORATION

Radar observations from the ground, supplemented by spectral and radio science measurements made aboard spacecraft, can allow an improvement to be achieved in the measurement of Mercury's static and dynamic figure, thereby providing additional (indirect) insight into the present composition, structure and (by implication) process of formation of this planet.

7.3   PLANNED MISSIONS AND THE INTERIOR

As indicated in Chapter 2, there are presently two orbital missions to Mercury with official status (namely NASA's Mercury MESSENGER and the ESA/ISAS Bepi Columbo mission) both of which could provide relevant measurements.

The MESSENGER mission to Mercury (MM) (**Figures 2.5 and 2.6**) was launched in August of 2004. Measurements from many MM instruments (**Table 2.3**), including X-ray, Gamma-ray, Neutron, and Visible and infrared spectrometers, along with the Laser Altimeter, Radio Science Package and Magnetometer, will all provide constraints on composition and structure of the interior. The spectrometers will constrain the model of formation, by providing the first direct major and residual element bulk abundance estimates, polar volatile material identification, and iron-bearing mineral bulk abundance estimates for the northern hemisphere (Gold et al, 2001; Solomon et al, 2001). Our knowledge of the interior structure and core characteristics will be enhanced by an improvement over Mariner 10 in the determination of the second degree magnetic and gravitational fields. Gravity field coefficients C_{20} and C_{22} should be determined to better than 1%. Of course, attempts will be made to unambiguously determine the existence of a liquid outer core, constrain its outer radius and viscosity by measuring the amplitude of Mercury's forced physical libration (Peale, 1976, 1988), and improving knowledge of its second order gravity field and obliquity. The forced physical libration creates an irregular rotation which can be extracted using radio science (gravity) and altimetry (topography) data. It should be possible to estimate these parameters with 10% accuracy, which is sufficient to verify the existence of a liquid core and constrain its structure (Solomon et al, 2001).

The Bepi Columbo (BC) mission (Anselmi and Scoon, 2001; Spohn et al, 2001), now scheduled for launch in 2014, is far more ambitious than Mercury Messenger (**Figures 2.7 and 2.8**). BC is designed to provide more comprehensive coverage and improved accuracy in the determination of compositional, structural, and magnetic parameters. Mercury Messenger consists of a single fixed body spacecraft in a nearly polar, elliptical orbit with periapsis at high northern altitude. Bepi consists of two orbiters which will be operate simultaneously: the planetary mapping orbiter, a 3-axis stabilized spacecraft with a nearly polar, elliptical orbit with periapsis near the anti-solar point; and, the magnetosphere mapping orbiter, a spin-stabilized spacecraft with a highly elliptical orbit. The BC planetary mapping orbiter has instrumentation similar to the MM spacecraft, minus the altimeter and magnetometer, and the magnetosphere mapping orbiter has a fields and particles package in addition to the magnetometer (See **Table 2.4**). The Bepi spacecraft and mission design would allow additional capabilities for the planetary mapping instruments, including nadir pointing, better geometries, and improved illumination for the spectrometers which require it. Both orbiters have instruments which will provide coverage of the magnetic field and magnetosphere, in more detail than Mercury Messenger. Bepi Colombo should be able to provide harmonic coefficients for the gravity field through degree and order 25. At the anticipated spatial resolution of 400 km, this will generate a free air gravity accuracy of 0.1mGal. As for moments of inertia calculations, the expected accuracy for C/MR^2 is better than 3×10^{-4} and for C_m/C is better than 0.05. This level of accuracy will allow verification of the existence and further characterization of a partially molten core. Despite improvements in constraints on internal structure, derived models will be subject to non-uniqueness of all single platform gravity field measurements (Spohn et al, 2001).

7.4 THE FUTURE EXPLORATION OF MERCURY'S INTERIOR

Only missions to Mercury can provide spectral and field measurements from which to derive essential parameters necessary for (a) accurate modeling of this planet's thermal evolution, geochemical differentiation and core formation and (b) constraining current models of the origin of the inner solar system.

Combined spacecraft measurements from visible, near infrared, and X-ray, gamma-ray, and neutron spectrometers on the two planned missions should enable a sophisticated interpretation of the bulk distribution of elements and minerals (Clark and McFadden, 2000) not attainable from earlier ground-based spectral measurements. Bulk elemental ratios (e.g. Fe/Mg, Fe/Si, Fe/S), would be particularly useful in providing reliable estimates of the mean FeO content of the silicate portion (mantle) of

Mercury (Clark and Trombka, 1997). For instance, high iron abundance combined with low iron mineral abundance would lead to the conclusion that iron metal is present and that the planet is highly reduced. Core and mantle formation processes, as well as interior structure can be inferred from combined geochemical and mineralogical signatures of major volcano-tectonic terranes (e.g., smooth or inter-crater plains). These signatures would also be indicators of the style of volcanism (e.g., alkali versus mafic versus aluminum rich basalts), which determines not only the initial composition, but the subsequent fractionation process for siderophile, refractory, and volatile elements.

Indeed, both Messenger and Bepi Colombo missions feature polar orbiters designed for planetology. Bepi Colombo represents an improvement by including an additional spinner with a different suite of instruments designed for studying the external fields and particles environment. Certainly, despite limitations in coverage or spatial resolution for each mission, compositional measurements from either mission could be combined to provide essential compositional measurements. Magnetic and gravimetric parameters derived from measurements made by these two missions will allow non-unique yet modestly improved gravity and magnetic field models, providing greater insight into the core origin and structure.

However, a sophisticated multi-platform, multi-encounter approach with suites of identically instrumented platforms, would have the capability of taking simultaneous 'snap shots' in polar and equatorial locations. Such an approach would immensely improve our determination of Mercury's magnetic field, along with its internal structure. Multi-platform fields and particles data would resolve the non-unique magnetic field determination problem by providing simultaneous, multi-point magnetic field observations. Simulation of such an encounter strategy utilizing, generalized, inverse techniques (Connerney, 1981; Connerney and Ness, 1988), and assuming those magnetospheric noise characteristics observed during the Mariner 10 flybys, has confirmed that such a multiple close encounter strategy will resolve, without ambiguity, the spherical harmonic coefficients of the internal magnetic field up to octupole, with partial knowledge of higher degree and order coefficients. Knowledge of the magnetic spectrum to this accuracy is sufficient to determine the radius of Mercury's dynamo (conducting core), and allow its comparison with other known dynamos. Simultaneous polar and equatorial flyovers would provide an improvement in parameter uncertainties between 2 to 3 orders of magnitude better than single spacecraft measurements.

Again, a multi-platform mission would be well suited to investigate the gravity field and rotational dynamics of Mercury, thereby determining the distribution of mass throughout the planet and possibly the dynamical coupling between Core and Mantle. Preliminary studies indicate that such a mission would be able to measure J_2, C_{20}, and C_{22} to between 0.5 and 5%, depending on the viewing geometry. The current best measurements of the quadrupole moments are in error (one sigma) by 50 % (see above), but even this accuracy, taken together with topographic data from radar ranging to Mercury, allows the placing of limits on Mercury's crustal thickness of 100-300 km (Anderson et al, 1996). These limits can be narrowed considerably utilizing the better measurements of the gravity field provided by such a mission.

7.5 NEED FOR FURTHER INVESTIGATION OF MERCURY'S SURFACE

The limitations in both coverage and resolution for our set of observations for Mercury's surface, has left plenty of room for improving our understanding of the surface of Mercury. The highest resolution images, with resolutions of 100 meters, are available for limited areas in only one hemisphere. Minimal stereo coverage provides very little understanding of the morphology of features and terranes. These limitations have prevented comprehensive understanding of the impact history and age of the surface, because the smallest craters cannot be observed. The current models for volcanic and tectonic processes remain hypothetical, because many of the features postulated would be too small or degraded to be observed. Post-Mariner 10 improvements in spectral coverage and resolution in not only the infrared and visible regions, but X-ray and Gamma-ray regions, are needed to expand our understanding of the composition of Mercury's surface and exosphere. Theoretical calculations have indicated that X-ray spectrometry would be capable of providing major constraints on models proposed for Mercury's origin and terrane formation based on bulk abundances of major elements and variation in these abundances among major terranes (Clark and Trombka, 1997). Gamma-ray spectrometry would indicate regional variations in iron and alkalis as indicated by K.

7.6 GROUND-BASED OBSERVATIONS FOR SURFACE EXPLORATION

Perhaps surprisingly, because Mercury is a challenging target to observe from the ground, a number of ground-based observation programs will continue to provide important measurements, until and in some cases after the arrival of missions to follow up Mariner 10 (**Figures 4.5 and 4.6**). Near to mid-infrared red spectrometer observations, and low resolution imaging for the previously unimaged hemisphere, will be made with instruments recently improved in sensitivity and spectral coverage, whenever observing opportunities become available. Such observations will constrain mineralogy and allow advance planning and efficient use of resources for deep space missions. Although neutron spectrometers on the orbital missions planned will be used to confirm the evidence of volatiles at the poles, radar observations, with the improvements in sensitivity and spatial coverage recently made possible by upgrades at Goldstone and Arecibo, will continue, as they have in the past (**Table 4.4**) to provide topography and surface property data which neither planned mission will provide. In fact, the completion of imaging of the Moon at analogous frequencies has been suggested for comparison to Mercury.

7.7   PLANNED MISSIONS AND THE SURFACE

Both MESSENGER and Bepi Colombo mission will provide coverage for the unimaged hemisphere and polar terrain, as well as far more comprehensive data on the distribution, morphology, and composition of typical terranes and features, using imagers, a wide range of spectrometers, ranging and radio science devices.

The MESSENGER Mission, despite the limitation of a fixed body spacecraft and a highly elliptical (200 to 440 by 15,000 km) orbit with a high northern latitude (60 to 70 degrees), near-terminator periapsis, is designed to provide surface measurements which will fill in the gaps in surface coverage which remained after the Mariner 10 mission, including the northern polar region and most of the unimaged hemisphere (Gold et al, 2001; Solomon et al, 2001). In that way, the mission will allow the major issues involving the nature and history of surface processes on Mercury to be addressed (Solomon et al, 2001).

Wide and narrow angle multispectral imagers with 12 filters on a pivot platform (Hawkins et al, 1997) will provide global image maps at an average 250 m/pixel resolution and global multispectral maps at an average 1km/pixel resolution. In addition, stereo image coverage with 250 m/pixel resolution is planned for the entire planet. In this case, the near-terminator periapsis provides an advantage by providing low sun-angle views, which emphasize morphology, at the highest resolutions. Selected northern latitude areas containing crater ejecta, volcanic flows, or tectonic features, will have higher resolution coverage to determine crustal stratigraphy as well as tectonic and volcanic emplacement mechanisms (Milkovich et al, 2002). Features such as volcanic domes and highly degraded ridges should be observable if resolutions of tens of meters are available, particularly when image data is combined with the laser altimeter northern hemisphere topography map.

A number of spectrometers will provide measurements of surface composition and bulk compositional constraints important in ascertaining Mercury's origin (Solomon et al, 2001). The controversy over the presence and form of iron in the crust should be resolved by combining measurements of the one micron band associated with iron-bearing minerals from the near infrared spectrometer data with measurements of major element abundances, including iron, from the combined X-ray, Gamma-ray, and neutron spectrometers. Measurements of iron, mafic (Mg, Ti), alkali (K), and refractory element (Si, Al, Ca) abundances combined with abundances of heavy, radioactive elements (Th, U) by the gamma-ray spectrometer, as well as volatiles (O and H by the neutron spectrometer, S by X-ray spectrometer) should clearly distinguish the most likely model for origin and history of the planet. The various models for planet formation and subsequent volcanic history predict distinctly different signatures for iron as well as refractory and volatile elemental abundances. Variation in mafic and residual elements at the tens to hundreds of kilometer scale provided by the high energy spectrometers in the northern hemisphere (Clark and Trombka, 1997) should allow the nature of volcanic processes and signatures of major terranes to be determined and confirm that the color boundaries observed in Mariner 10 images are geochemical in nature. At that scale, observation of underlying stratigraphy overturned on the surface should be possible at larger craters.

The launch for the ESA/ISAS Cornerstone Bepi Colombo mission should occur after Messenger arrives at Mercury and thus is designed to provide surface observations which are both complimentary and more detailed than comparable observations from Messenger, which will certainly help in the planning of Bepi Colombo. Certainly the three-axis stabilized and nadir-pointing capability of the Mercury Planetary Orbiter (MPO) portion of this mission (Bepi Colombo Scientific Advisory Group, 2000; Grard and Balogh, 2001; Grard et al, 2002) is an advantage not available for Messenger. Another advantage for more uniform coverage of the entire surface will be the somewhat less elliptical 400 by 1500 kilometer orbit. Such coverage should allow discovery or confirmation of systematic variations in regolith properties, such as variations relative to Mercury's two 'hot poles' induced by differences in the degree of thermal annealing and rates of micrometeoroid impact and radiation damage which result from the intense interaction between surface grains and the particle flux or solar wind at Mercury's location (Langevin, 1997).

The MPO will carry two imaging cameras. The wide angle camera resolution is limited to only 200 m or better, limited by available data volume. For selected areas, accounting for 5% of the surface, 20 m resolution will be available with the narrow angle camera. The Messenger mission can act as a precursor, enabling the areas of greatest interest to be identified and targeted for high resolution viewing (Grard et al, 2001).

A wide range of spectrometers includes IR, UV, X-ray, Gamma-ray, and neutron detectors. Measurements from these spectrometers will allow more consistent determination of the bulk, major terrane, and typical feature components, including abundances of major elements and minerals, and volatiles, including water ice. Major elemental abundances at resolutions of kilometers to tens of kilometers will be provided by an X-ray spectrometer on Bepi Colombo. and radioactive trace element (K, Th, U), high cross-section element (Fe, Ti), and volatile (H,O) abundances will be provided by combined gamma-ray and neutron spectrometer measurements at hundreds of kilometers resolution, just as for Messenger. The inherently higher spectral resolution and sensitivity of the proposed solid state X-ray detector (SMART1/CIXS, Grande et al, 2001) should allow quantitative determination of a greater number of lines with higher sensitivity, including Na, Mg, Al, and Si, as well as K, Ca, and Fe during solar flares. Measurement of bulk K/U or K/Th ratios necessary for understanding Mercury's thermal history will certainly be possible with a state of the art gamma-ray spectrometer (Bruckner and Masarik, 1997). Combined measurements would provide further constraints and convergence on a working model for Mercury's formation. Considerably more information on mineralogy will be provided by the infrared spectrometer in the 0.8 to 2.8 micron region, a much broader region than available with Messenger, at a far better spectral resolution of better than .01 microns.

The symmetrical orbit of the nadir-pointing MPO will allow more comprehensive and uniform mapping of surface morphology and interior structures from which the surface features are derived from combined observations made by the radio science, ranging, and magnetometer experiments.

7.8   THE FUTURE EXPLORATION OF MERCURY'S SURFACE

A near-polar orbiter mission to Mercury is the only way to provide spectral and physical observations of the surface with the necessary resolution for breakthroughs in our understanding of tectonism, volcanism, and impact, thereby allowing the resolution of controversies concerning the timing of major historical events on Mercury. Both Messenger and Bepi Colombo have instruments to provide such measurements.

Of course, an ideal outcome would be coverage for the unimaged (by Mariner 10) hemisphere, as well as more uniform coverage of the entire surface, at the highest resolution possible. Especially optimal would be timing arrivals so that the sub-solar point would be greatly offset in longitude from subsolar longitudes of Mariner 10, allowing complementary viewing of many features (lower sun angle coverage for features that had higher sun angle coverage on Mariner, and vice versa). Global imaging must be of sufficient resolution to provide global inventories of volcanic source areas such as vents, cones, or shields (which give essential clues to variations in composition, internal structure and temperature), as well as of features representing major tectonic episodes such as compressional scarps (from which the extent and history of crust/mantle/core interaction and core formation can be determined). Global determination of depth/diameter ratios for the craters in the ten's of meters size range, unseen by Mariner 10, would allow age determinations to be made over Mercury's entire surface (Strom and Neukum, 1988). Combined close range observations of these features made by X-ray, Gamma-ray, and Neutron spectrometers could allow determination of distinctive geochemical signatures, and thus origins, for typical features and terranes, and of underlying strata exposed at the surface at impact sites.

Both planned missions have some limitations in either resolution or coverage or both. Due to its high northern latitude periapsis and lack of nadir pointing capability, Messenger will provide poor resolution coverage by spectrometers and imagers for the southern hemisphere. The Bepi Colombo planetology mapper does not have these limitations, and thus will provide more uniform coverage of northern and southern hemispheres; however, its farside periapsis means high altitude over the illuminated hemisphere and thus poorer spatial resolution, no better than 200 meters for images, for most of the surface. Fortunately, for both missions, images with ten times better resolutions can be provided for 5% or less of the surface. Recordable bits/pixel are limited in both missions due to limitations in allocation for mass, and thus data volume, for onboard data storage and for download bandwidth.

How can we fly a Mercury mission which would allow for higher resolution coverage for more of the surface? Optimally, a polar orbiter would have near-subsolar, equatorial, and relatively low altitude periapsis and would not have limitations in data volume. One way is to wait until funding or serendipity provides major breakthroughs in thermally resistant materials, and this allows a mission with a low altitude, nearly circular orbit to be flown. However, another entirely different strategy, which could be used now, could involve using a multiple encounter flyby with a solar orbital period designed for just such a close encounter with Mercury every other Mercury year when alternate hemispheres are illuminated. Flybys can have much larger scientific payloads because of far lower onboard fuel requirement. Thus, they can provide for much greater data storage and download over much longer periods.

7.9   NEED FOR FURTHER INVESTIGATION OF MERCURY'S EXOSPHERE

The bulk abundances and distribution of most species in Mercury's atmosphere are still relatively unknown (Killen and Ip, 1999). Even for those that are relatively well known, Na and K, the abundance ratios show tremendous variation, for reasons that are not completely clear (Potter et al, 2002).

There is still strong disagreement as to the relative importance of the individual processes that supply and remove exospheric material at Mercury. (see for example McGrath et al. 1986, Cheng et al. 1987 Sprague 1990 and Morgan and Killen 1997). Sputtering can, for instance, inject all the common regolith species into the exosphere (O, Si, Ca, Al. Mg and Fe). Impact ionization preferentially supplies volatiles (S, H_2O and OH) in addition to regolith species (Killen et al. 1997). Crustal diffusion has been predicted to provide regolith derived species to the exosphere (Sprague, 1990).

The extent to which exospheric species abundances are diagnostic of the composition of surface rocks, and the nature of interaction between the solar wind and the magnetosphere are thus important questions which remain to be answered (Killen et al, 2004).

The fate of photo-ions on the nightside, whether accelerated down the tail and lost or adsorbed on the surface and retained, is an unresolved issue which has major implications for understanding the evolution of the atmosphere (Killen et al, 2004).

7.10   GROUND-BASED OBSERVATIONS AND THE EXOSPHERE

Ground-based observations will continue to provide higher resolution spatial and spectral measurements of Na and K, and more detailed emission feature maps for Mercury as the spectrometers improve in sensitivity.

However, a major thrust of the ground-based observation program will be more comprehensive modeling of atmospheric species sources and sinks, and their interactions with the surface and magnetosphere (Killen and Ip, 1999; Killen et al, 2004). Better determination of ratios of measured constituents (e.g., Na/K) will give important clues to the origin of the atmosphere. Particular areas of interaction being considered are 1) the relationship between atmospheric and surface composition, 2) the dynamics of the atmosphere, 3) interaction between the magnetosphere, solar wind, and atmosphere.

7.11   PLANNED MISSIONS AND THE EXOSPHERE

Of particular significance for the planned missions to Mercury is the measurement of those species which could have major implications for understanding the influence of surface composition (Ca), and the nature of volatiles in the crust (S or OH), but can't be effectively measured from the Earth, due to the difficulty of observing at the UV wavelengths of their characteristic lines or their relatively low abundances (Killen and Ip, 2004).

The Messenger Mission (Solomon et al, 2001) will fly one instrument dedicated to measuring atmospheric species, and several others which will provide important data on the interaction of atmospheric species with the surface and magnetosphere (Killen et al, 2004). Such measurements will allow dominant source mechanisms to be quantified and also provide insight into the composition and distribution of volatiles in upper crustal layers.

The Mercury Atmospheric and Surface Composition Spectrometer (MACS) will measure the composition and structure of Mercury/s exosphere; study its neutral and coronal gas and measure ionized atmospheric species. This instrument combines UV/visible and UV/IR spectrometers, providing a comprehensive inventory of atmospheric species and determining their spatial and temporal distributions. Spectra from 0.115 to 0.600 microns at 1 nm resolution will provide altitude profiles of known species (H, Na, K and Ca) and to search for predicted species not previously detected (Si, Al, Mg, Fe ,S, OH) as well as new species. Limb scans will be made by 'nodding' the spacecraft to provide altitude profiles of emission lines. (Solomon et al. 2001).

In addition, four other spectrometers will provide compositional information on the surface. The Visible-Infrared (VIRS) spectrometer will measure reflectance from 0.3-1.4 micrometers with a spatial resolution of approximately 5 km. This wavelength range contains spectral signatures of Fe-bearing Silicates and Ti-bearing minerals. The X-ray and Gamma-ray spectrometers will provide measurements of elemental abundances for Mg, Al, Si, Ca, K, and Fe. The neutron spectrometer will provide measurements of hydrogen, from which water, or ice abundance at the poles, could be inferred. These measurements will be combined with multi-spectral images obtained with the Mercury Dual Imaging System to study the effects of space weathering at the surface and potential sources for exospheric material. Combined observations will contribute to our understanding of the processes that generate and maintain the atmosphere; elucidate the connection between atmospheric and surface composition and the transport of volatile materials on and near Mercury and determine the nature of the bright radar reflective material at its poles (Solomon et al, 2001).

The Energetic Particle and Plasma Spectrometer (EPPS), described below in the Magnetosphere section, will provide complementary measurements of plasma composition. Plasma composition is important because of the close coupling between Mercury's surface, exosphere and magnetosphere. In addition, species in the exosphere-magnetosphere system are diagnostic of volatiles present on the surface (Cheng et al, 1987).

The payload proposed for the Bepi Colombo MPO is similar to the payload already flying on MESSENGER. A UV spectrometer will provide direct measurements of atmospheric species which cannot be observed from the Earth. This instrument will be complemented by spectrometers operating in the infrared and visible regions which will provide mineralogical measurements of the surface and emission line measurements from the limbs, as well as X-ray, Gamma-ray, and neutron spectrometers providing elemental abundance measurements of the regolith and proton (water) abundances at the poles. Such a payload will enhance understanding of the spatial and temporal distribution of atmospheric species, as well as the interaction between exosphere and surface. In addition, a neutral and ionized particle analyzer will be on board, allowing characterization of particles that interact with atmospheric as well as surface constituents, resulting in a process known as space weathering. Combined simultaneous measurements from these instruments should allow great advancement in our understanding of atmospheric processes, sources, and sinks.

Bepi Colombo proposed orbital trajectories are quite different from MESSENGER orbital trajectories. Bepi Colombo MPO periapsis is at the subsolar point but at a higher altitude, allowing the atmosphere to be observed under different conditions. As a result, further characterization of suggested east/west and north/south asymmetries in the distribution of atmospheric species should be possible.

7.12   THE FUTURE EXPLORATION OF MERCURY'S EXOSPHERE

Close range observations of Mercury will provide a far more comprehensive survey of Mercury's atmosphere than Earth-based observations can provide. With the exceptions of Na and K, Mercury atmospheric components cannot be observed through Earth's atmosphere with ground-based telescopes because their emissions occur in the UV. Earth-based observations have yielded low resolution information on spatial and temporal distribution of these species. In the future, space-based telescopes, which could get around this restriction imposed by the Earth's atmosphere, will still not be allowed to observe Mercury because of constraints on pointing close to the sun.

UV/visible spectrometer measurements combined with measurements from surface spectrometers onboard both Messenger and Bepi Colombo will identify all the atmospheric species data needed to determine the present nature of the exosphere, and far greater understanding of its origin, and the role of atmosphere-surface interactions in its formation. Additional information on atmosphere/magnetosphere coupling; and thus a more comprehensive model of exosphere processes, could be provided by the Bepi Colombo suite of particle and plasma detectors on the magnetospheric orbiter, particularly if its orbit remains in resonance with its planetology orbiter.

However, the atmosphere is a dynamic system, with variations on the short time scale associated with magnetospheric and solar wind variations. Just as in the case with the magnetosphere, and interior core/magnetic field interactions, this system can only be understood by getting a three dimensional snapshot, from simultaneous measurements obtained from suites of identical instruments on multiple platforms placed in key locations at poles, subsolar and anti-solar points, terminators. Neither present mission allows this possibility. Such instrument suites would include UV/Visible spectrometers for lines of atmospheric species along with instruments which would measure magnetospheric and solar wind variations as described in the next section.

7.13 NEED FOR FURTHER INVESTIGATION OF MERCURY'S MAGNETOSPHERE

Further investigation of Mercury's magnetosphere is particularly needed to address the issues of Mercury's magnetosphere similarities to and differences from the Earth's. Mercury is the only other terrestrial planet to have an analogous magnetosphere (Slavin, 2004). Because Mercury's proximity to the sun and relatively small size and gravity result in more rapid, frequent, and intense interactions between the solar wind, magnetosphere, and surface, the planet represents a wonderful opportunity to study magnetosphere dynamics. Of particular interest is an understanding of the nature of the solar wind/magnetosphere interaction and magnetopause reconnection.

Knowledge is particularly sought on the manner and timescale in which the solar wind enters the magnetosphere and its impact on surface and atmosphere components. Energetic particle composition, sources and sinks, and populations at equilibrium as well as during acceleration and transport, and even acceleration processes, are not well known for Mercury. Tremendous variation in the temporal and spatial distribution of exospheric constituents have been observed from ground-based telescopes (Killen and Ip, 1999). Models indicate that neutral particles could be entering the atmosphere through sputtering induced by solar wind and magnetosphere ions, or through less energetic interactions, but systematic observational coverage is difficult to obtain and thus lacking. The fate of such ions and their influence on magnetospheric processes such as reconnection, convection, and wave propagation, is still poorly understood (Slavin, 2004). Current-related processes are affected by the nature of coupling between the magnetosphere and the planetary magnetic field on Mercury as on the Earth (Slavin, 2004). Electrodynamics of the interactions between magnetosphere, solar wind, and the surface, are poorly constrained due not only to the limited nature of in situ measurements available. Mercury's magnetosphere, unlike the Earth's, can interact directly with the regolith, because Mercury lacks on ionosphere (Killen et al, 2004). Thus, processes including field-aligned current closure, induction current generation, substorm generation, and magnetotail reconnection, are relatively poorly understood (Slavin, 2004).

Ongoing work on the development of magnetospheric models which constrain interpretations of observed exosphere distribution and vice versa is particularly relevant (Killen et al, 2004). A magnetosphere of considerable complexity, interacting with surface and atmosphere, is emerging.

7.14   GROUND-BASED OBSERVATIONS FOR MAGNETOSPHERE EXPLORATION

The study of Mercury's magnetosphere requires in situ observations that must be provided by spacecraft. However, considerable effort has been addressed to the developing and models of Mercury's magnetosphere and using them to constrain interpretations of atmosphere and surface observations.

7.15   PLANNED MISSIONS AND THE MAGNETOSPHERE

On MESSENGER, the EPPS (Energetic Particle and Plasma Spectrometer) will provide measurements of the composition, distribution, and energy of electrons and various ions in Mercury's magnetosphere. Both planetary and solar wind ions must be present at the Bow Shock, Magnetopause and Cusps and in the plasma sheet. The connection between exosphere and magnetosphere will be achieved through comparing measurements of energetic ions, electrons and thermal plasma ions and atmospheric species measurements.

In addition, the magnetometer on MESSENGER will characterize Mercury's magnetic field in detail from orbit over four Mercurial years. In addition, the magnetometer on MESSENGER will characterize Mercury's magnetic field in detail from orbit over four Mercurial years. Combined magnetometer and EPPS measurements will allow the effects of the Sun on magnetic field dynamics to be characterized to a greater extent than previously possible.

Bepi Colombo MMO has a much larger suite of instruments to characterize Mercury's magnetosphere and its interactions with the exosphere and surface, including a magnetometer similar to the one flying on MESSENGER, allowing direct measurement of magnetic field components. Particle analyzers for ions and electrons over a wide range of energies, and for energetic neutrals, should allow extensive characterization of the distribution of particles in the magnetosphere. Electric Field, Electric and Radio Wave instruments should allow characterization of the structure, sinks, and sources of Mercury's magnetosphere. Two complementary instruments provide imaging of Na and detection of dust, which should greatly advance our understanding of the interaction of solid particles and volatile (exospheric) components with charged particles in the magnetosphere.

7.16 THE FUTURE EXPLORATION OF MERCURY'S MAGNETOSPHERE

The magnetosphere and its interactions with surface, exosphere, and solar wind, can only be studied at close range. The two planned missions to Mercury, Messenger and Bepi Colombo, will provide information which we don't have now on Mercury's magnetosphere. Both carry compositional spectrometers and magnetometers on near-polar orbiters, to provide information on the magnetic core, interior, surface, and atmosphere. Messenger is really a planetology mission, and carries one instrument designed to measure the external particle environment, the high energy particle detector, and thus will not provide very comprehensive information on the magnetosphere. Bepi Colombo has a separate magnetospheric orbiter, with a comprehensive suite of magnetospheric particle, plasma, and field instruments to provide measurements on the structure of the magnetosphere and its interaction with the solar wind at low spatial and temporal resolution.

However, only multi-platform, simultaneous spacecraft observations at poles and equator could determine the 3D structure of the Hermean magnetosphere and its interaction with the solar wind as well as the nature of magnetospheric substorms by combining high time resolution, multi-point magnetic field measurements and simultaneous wave, and particle measurements with state-of-the-art modeling codes to develop a global picture of Mercury's magnetosphere and its key boundaries. A minimum of three platforms would allow determination of three pairs of nearly simultaneous, magnetopause locations (which would provide the 'best fit' magnetopause shape). These data could reveal both equator-pole asymmetries and asymmetries due to reconnection, as are observed at the Earth (Petrinec and Russell, 1993). Such a mission would have the advantage of beginning a modeling effort after equivalent codes for the Earth's magnetosphere have already attained a high level of maturity, sophistication, and power. Numerous data-model comparisons have clearly demonstrated the value of such models in determining physical context from multi-point measurements. Thus, the ability to simultaneously measure both polar caps will unambiguously resolve several questions directly applicable to terrestrial magnetosphere studies. Spacecraft must be instrumented to determine the efficiency of IMF reconnection; derive from particle and field measurements the size, total magnetic flux, and asymmetry of the caps; measure the cross-polar cap potential drop to yield the overall strength of Mercurian convection (Boyle et al., 1997), a key parameter in determining whether observed substorm like phenomena at Mercury are driven by reconnection. In turn, these parameters would normalize the models, allowing predictions of the size, shape, and orientation of the magnetic field patterns for other IMF orientations. A (single) orbiter could not directly measure the amount of magnetic flux connecting to the solar wind, nor the flux of impacting solar wind.

The importance of induced magnetic fields for the structure and dynamics of the Hermean magnetosphere has been widely discussed (Hood and Schubert, 1979; Suess and Goldstein, 1979; Glassmeier, 1999). The simulataneous multi-platform magnetic meaurements taken near Mercury would allow the solar wind field, and the induced field to be separately measured. Also, the electrical conductivity structure of the planet could be deduced from the magnetic measurements. A simple ENA Imager (Barabash et al., 1997) could uniquely provide important information regarding the dynamics of the magnetosphere namely (1) the frequency and duration of plasma injections (substorms); (2) typical decay times of the substorms; (3) energies of the stimulating plasma injections (via ENA spectrum measurements). Since ENA measurements can be conducted remotely while the spacecraft is approaching/receding from the planet, a significant increase in the total period of measurement relative to the flyover time can be achieved. Moreover, ENA images showing instantaneous global pictures of ions distributions are inherently free form the ambiguity of spatial-temporal variations effecting in situ measurements. ENA measurements could also reveal the mass composition of the magnetospheric plasma.

Ionization of the exosphere would create low energy plasma detectable by properly instrumented platforms. Solar wind induced electric fields either accelerate ions into the magnetosphere or back to the surface where they are reimplanted. In the extreme case when solar wind impinges directly on the surface, the ions would gain energy at a rate of a few eV/ km (resulting in energies in the keV range for ions created near the surface). Regions of ion charge exchange and pickup would be defined through observations of low frequency hydromagnetic waves detectable as long-period magneto-sonic fluctuations (e.g., Richardson et al., 1987).

Because the solar output is more intense at Mercury than elsewhere in the solar system, there is a continuous interaction of energetic particles with the solid surface of the planet. These interactions continuously process the surface and the sputter products they produce, in turn, feed the exosphere and magnetosphere. These particles can, thereafter, be accelerated in the magnetosphere and the cycle continues. Observations taken aboard Mariner 10 suggested that the auroral zone, where the energetic radiation interacts with the surface, is a potential source of intense X-rays (Grande, 1997). The solar wind may also contribute to X-ray generation, particularly if it can reach the surface during highly excited periods. The correlation of ENA, X-ray and energetic particle observations can allow the characterization of Hermean global energetic and magnetospheric dynamics. Correlation of energetic particle circumstances in the solar wind (inbound and outbound) with ENA and X-ray observations, can provide special insights into the response of Mercury's magnetosphere to different kinds of interplanetary activity. From such combined data, it would be possible to determine the density and composition of the magnetospheric ionized and neutral components, as well as the flow and temperature of the ionized components. Also, electron and ion distribution functions will be determined. The identification in magnetic data of predicted quasi-periodic effects associated with ionosphere-magnetosphere coupling and substorm development (Glassmeier, 1997), can profoundly influence the terms of the current debate in the literature concerning the nature of Hermean substorms. Multidisciplinary magnetospheric data from Mercury would have a unique role to play in the discipline of comparative magnetospheres.

7.17   CONCLUSIONS: A NEW APPROACH

The capabilities and limitations missions Messenger and Bepi Colombo have in meeting the next generation of exploration goals is summarized in **Table 7.1**. Orbital missions have several disadvantages, including large onboard fuel requirement and enhanced thermal shielding requirements which compromise payload size and operation.

A low-cost, express, multi-platform, multi-flyby, designed to encounter alternately illuminated hemispheres (Clark et al, 2002) could 'fill in the gaps' in several ways. First of all, a flyby requires little in the way of

Table 7-1. **Past, Planned, and Proposed Mission Measurement Goals.**

Measurement	M10 1974	MM 2011	BC 2017	Proposed
Bulk Composition/Inner Structure	0	2	2	2
Geological Survey	1	1	2	2
Mineralogical Survey	0	1	2	2
Atmosphere Survey	1	1	2	3
Atmosphere/Magnetosphere Interaction	0	1	2	3
Core/Mantle/Surface Interaction	1	1	1	2
3D Magnetic Field Modeling	0	0	1	3
2nd Order Gravitational Modeling	0	1	1	2
Magnetosphere Structure	1	1	2	3
Magnetosphere Composition	0	1	2	3
Plasma source/Acceleration	1	1	2	3
Mission: M10 Mariner 10 MM Messenger BC Bepi Colombo				
Exploration: 0 No 1 Partial 2 Complete Primary 3 High Order				

Table 7-2. **Science Goals Still to be Met**

System	Science Objectives	Measurement Goals
Interior	1) Determine source, strength, and orientation of Mercury's internal magnetic and gravitational fields. 2) Determine core/mantle/ crust interaction	1) Obtain direct measurements of low altitude, external magnetic field near equator and both poles. 2) Determine moments of inertia C_{20} and C_{202} to better than 10% accuracy. 3) Determine internal moments of magnetic field to 3rd order.
Surface	Determine nature of crustal rocks and minerals and the history of tectonic, volcanic, and impact features and typical terrane signatures at highest resolution	Obtain high resolution spectral, imaging coverage for representative tectonic, volcanic, and impact features.
Atmosphere	1) Determine spatial and temporal distribution of atmospheric constituents. 2) Understand interaction between exosphere, magnetosphere, solar wind, surface, interior.	1) Determine distribution of atmospheric constituents. 2) Measure distribution of atmospheric ion species in solar wind plasma, magnetosheath, and magnetosphere.
Magnetosphere	1) map and model magnetosphere in 3D. 2) Determine sources and sinks and bulk composition of magnetospheric plasma. 3) Determine interaction of solar wind, exosphere, and surface with the magnetosphere.	1) Obtain 3D magnetic (to 3rd order) and gravitation (to 2nd order) field measurements. 2) Monitor magnetospheric Dynamics at high time Resolution. 3) Obtain high spatial and temporal resolution particle, wave, and magnetic field measurements.

expendables, which would allow for a large scientific payload to be launched. A multi-platform mission with identically instrumented probes capable of taking simultaneous measurements in the vicinity of Mercury could generate a 3D snapshot which conventional orbiters could not provide. Such a mission would need to meet the following essential science objectives: (1) a comprehensive survey of Mercury's interior, surface, atmosphere, and magnetosphere; (2) a detailed determination of the 3D magnetic structure of Mercury's magnetosphere and its interaction with the solar wind; (3) unique, high time resolution, monitoring of magnetospheric dynamics; (4) Systematic coverage at highest resolutions to allow direct comparison of east and west, north and south hemispheres, from which essential parameters for models of core formation and interior differentiation processes can be derived, as well as (5) a global survey of Mercury's major terranes (tectonic, volcanic, and impact structures); and (6) tests of gravitation theory. Future science goals are summarized in **Table 7.2**.

Our Express mission concept calls for a multi-platform, very fast, high-return mission to Mercury requiring no advances in technology, having a launch window approximately once a year, and with the capability of being launched in well under 5 years. The main spacecraft would include a full complement of instruments, as described below, as well as two to three additional miniprobes with suites of 3 to 4 fields and particles instruments identical to those flying on the main spacecraft. The spacecraft would have a direct ballistic trajectory to Mercury allowing arrival at the target 100 days after launch. The miniprobes would be passively launched with trajectories over the poles and through the magnetotail just before flyby. Simultaneous measurements by the three to four fields and particles instrument suites would allow the capture of a 3D snapshot of Mercury's field and particle environment, and determination of gravity and magnetic fields to very high degree and order. One maneuver would then place the spacecraft into a solar orbit, allowing a second encounter with the opposite illuminated hemisphere 264 days later, thus completing the second flyby approximately one year from launch. This approach minimizes risk from exposure to thermal radiation by minimizing time spent close to the sun. The mission approach is illustrated in **Figures 7.1, 7.2, and 7.3**.

Minimum fuel requirement would translate into a large payload, allowing high capacity onboard memory. High data volume could be burst into high capacity onboard memory at each encounter, or playback as convenient, at a low rate after each encounter.

Details of the payload on our proposed multi-platform, multi-flyby mission is summarized in **Table 7.3**.

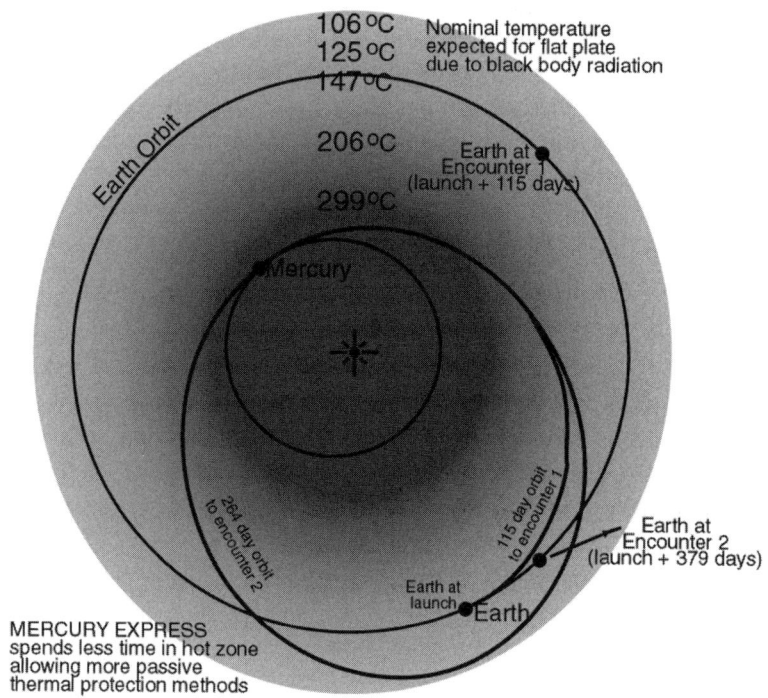

Figure 7-1. Trajectory of Proposed Multi-Platform Express Mission in solar equatorial plane illustrating minimal exposure to thermal radiation.

Planetology Instruments: By providing bulk abundances of major elements, an **X-ray Spectrometer** would determine which model of Solar System formation is most valid. **A Visible/Near Infrared Imaging Spectrometer** would determine the abundances of Fe- and Ti-bearing minerals. When data from these spectrometers are combined, a sophisticated interpretation of the petrology of major surface features, not possible with earlier spectral measurements, can be made (Clark and McFadden, 2000). Core formation and interior structure can be inferred from geochemical and mineralogical signatures for major volcano-tectonic terranes. These signatures also indicate the style of volcanism (e.g. alkali, mafic, aluminum rich basalts), thereby determining not only the initial composition, but also the subsequent fractionation process for siderophile, refractory, and volatile elements. A **Visible/Near Infrared Imaging Spectrometer** would provide a global geological inventory of Mercury (leading to breakthroughs in our understanding of tectonism, volcanism, and impact activity) as well as insights into the timing of Hermean geological events. **Magnetometer** data would resolve the non-unique, global, magnetic field determination problem through its simultaneous, multi-point magnetic field measurements (spherical harmonic coefficients of the internal magnetic field, to octupole).

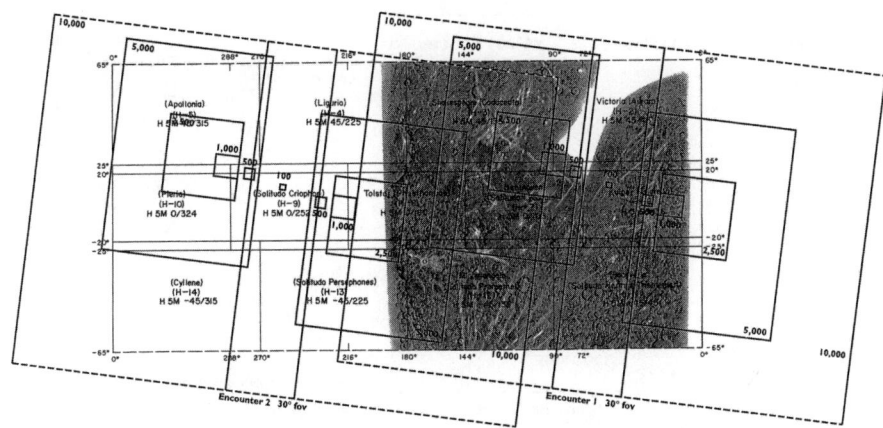

Figure 7-2. Along-Track Footprints for Proposed Mission encounters of alternate hemispheres from perspective of nadir-pointing for an spectrometer with a wide angle field of view. Darker gray dashed lines represent the terminators during each encounter.

Resulting knowledge of the magnetic spectrum would be sufficient to determine the radius of the dynamo (conducting core) and allow comparison with other known dynamos. A **Gravity Experiment** would determine C_{20} and C_{22} to a 0.5 - 5% accuracy, i.e. a 1-2 order of magnitude improvement over present estimates (Anderson et al, 1988, 1996). These observations, combined with an estimate of Mercury's physical libration and obliquity, would provide a direct measure of Core-Mantle coupling (Peale, 1976, 1988) and enable estimates of Mercury's crustal thickness.

Atmospheric/Magnetospheric Instruments: An **UltraViolet Spectrometer** would provide a comprehensive atmospheric survey and also atmospheric species maps. The surface/atmosphere/magnetosphere/solar wind interaction can be derived when data from the planetology spectrometers are combined with particles and fields data. A **Low Energy Plasma Analyzer** would provide a direct determination of the composition of the solar wind particles. An **Electric Field Instrument** would make an independent determination of electron density and temperature using thermal noise analysis, (Meyer-Vernet and Perche, 1989) (independent of spacecraft charging). The ions created by charge exchange with magnetospheric and solar wind ions excite low frequency (sub-Hz) hydromagnetic waves detectable by **Magnetometer** as long-period magneto-sonic, fluctuations. At sufficiently high ion densities, higher frequency ($f < f_c$) ion acoustic and whistler mode waves will become important (Richardson, 1987) and could be observed by **Search Coil Antenna** and **Electric Field Instruments** (these waves are produced by ring-like and beam-like distributions of ions).

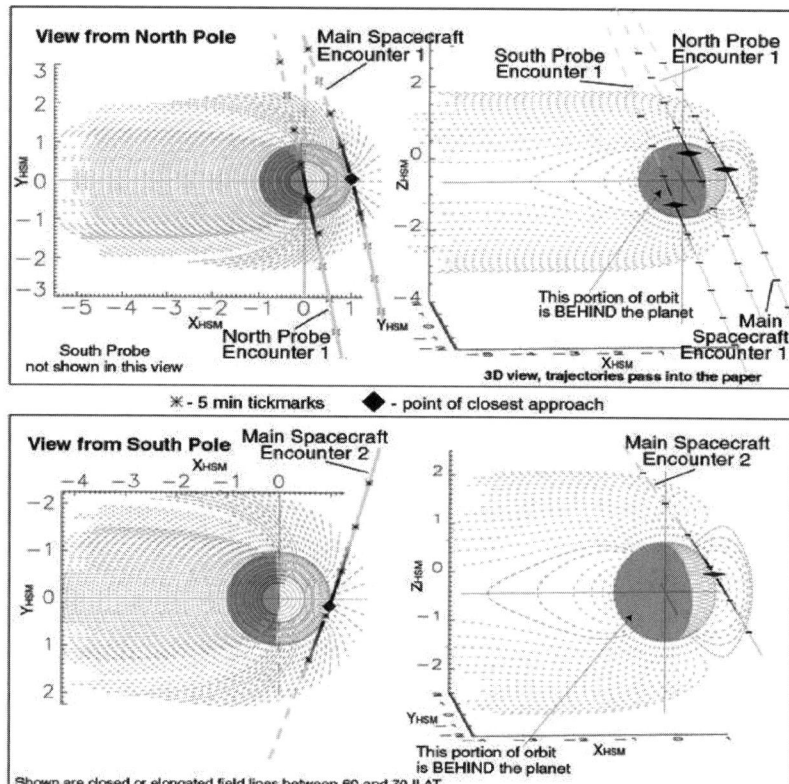

Figure 7-3. Simultaneous Paths through Magnetosphere for Probes. Orientation of Multi-platform express spacecraft trajectories through model Mercury magnetosphere and along ground track.

Measurements of the wave modes would provide high time resolution information in regard to particle distributions in the magnetosphere, facilitating insights into dynamic interactions between Mercury's magnetosphere and atmosphere and allow sensing of bow shock transitions. **Energetic Neutral Atom Analyzer** with **Low Energy Plasma** and **Energetic Particle Spectrometers** would determine the density and composition of the ionized and neutral components, the bulk flow and temperature of the ionized components, and electron and ion distribution functions. Just prior to each Encounter, measurements of EM waves measured by an **Electric Field Instrument** and **Search Coil Analyzer**, generated at twice the local plasma frequency (2fp ~ 120kHz) and freely propagating away from the foreshock region, would signal approach to the Bow Shock. Measurements of the IMF direction by a **Magnetometer**, and of wave activity and of enhanced fluxes of backstreaming electrons by a **Low**

Energy Plasma and **Energetic Particle Spectrometers,** would define the leading edge of the foreshock and the location and general shape of the bow shock itself. An **Energetic Neutral Atom Analyzer** would help to characterize the nature of substorm activity in the highly variable magnetospheric environment. An **Energetic Particle Analyzer** would monitor the response of the magnetosphere to different kinds of IP activity and determine the penetration of IP ions and electrons into the magnetosphere.

Table 7-3. **Future Multi-platform Mission Instrument Suite**

Instrument	Measurement	Data Products
*Visible/Near IR imaging spectrometer	Images accompanying spectra, .4-2.5 u, .05 u resolution	Pyroxene, Olivine, Ilmenite abundance, Ca, Al abundance at limbs, tens meter resolution along track
*Vis/UV spectrometer, 5 kg	Spectra, 700-3300 A, 10A resolution	Temporal, spatial maps all atmospheric species
*X-ray spectrometer	Spectra, 1-10 keV, 200 eV resolution,	Major element abundances Tend km resolution along track
Magnetometer	DC magnetic field, +/- 655 nT resolution, 0-1 hz noise	3D magnetic field components, anomaly map, core formation models
Energetic Particle Spectrometer	Electrons .01-1 MeV, Protons .025-50 MeV, 10msec resolution, 360 degree coverage, 15 degree pitch angle bins	Plasma moments, flux intensity and ion spectra distribution in space and time, radiation belts, magnetosphere structure
Low E Plasma Detector	3-30000 eV, 3sec/3D scan, 1-135 amu/Q, ions, electrons	3D distributions, plasma moments, ion flux models
Energetic Neutral Atom Imager	Images, 10-1000 keV, 30% E and <20 second time resolution	Frequency and duration of plasma injections and decay times of storms
Search Coil	AC magnetic field spectrograms, 3 spectra/sec, .01-10 kHz	Wave forms, frequency/time spectrograms, 3D magnetosphere structure
Electric Field	AC electric field spectrograms, 3 spectra/sec, 1-500 kHz, 64 channels	E field and electron density plots, spectrograms, source direction, radio intensity
Transponder, radio science	Coherent X-band, doppler/range spectra, gravity field anomalies	J_2, C_{20}, C_{22}, 2nd order gravity maps, core/mantle coupling and crustal equilibrium
* Subsolar equatorial periapsis platform only		

The methodology and technology for creating identically instrumented multi-platform missions is currently being developed and applied to terrestrial magnetospheric missions. Such a multi-platform mission to Mercury would not require one of the major technological developments required for these missions, where careful control or assessment of position within a formation is required. The biggest challenges would be building a well-calibrated instrument suite, requiring proper allocation of resources and facilities, a capability now being developed for the MMS mission (Curtis et al, 2004).

The concept for such a mission would fully exploit the ability of an inexpensive launch vehicle to directly transfer a payload from the Earth to Mercury in just over 100 days and return to Mercury for a second encounter 264 days later, when the planet's opposite hemisphere is illuminated. The resonance between Mercury's orbital and rotational periods is utilized to minimize the time between encounters with opposite hemispheres (<9 months) and thus to minimize mission duration (thereby limiting the thermal radiation and onboard propulsion requirements associated with long flight times). These trajectories would be comfortably achievable by either the Delta II 7925H or the Soyuz equivalent launch vehicles and there are two-week launch windows every 340 days. The minimum C^3 (twice the combined kinetic and potential energies per unit mass in km^2/sec^2 required for the launch vehicle) and ΔV (the velocity change required from the spacecraft propulsion system in m/s to achieve the heliocentric period change to ~264 days) for up-coming launch opportunities are shown in **Table 7.4** Thus, such a mission could be flown today, to provide a truly comprehensive view of Mercury in the context of its environment.

Table 7-4. **Typical Opportunities for Proposed Multi-platform Flyby**

Launch	Arrival	C3	Delta V
12/20/05	04/01/06	44.7	902
12/01/06	03/16/07	42.0	777
11/12/07	03/01/08	40.7	808
10/25/08	02/16/09	41.5	1179
10/10/09	02/02/10	44.7	1533

7.18 REFERENCES

Acuna MH, J.E.P. Connerney, P. Wasilewski, R.P. Lin, J.A. Anderson, C.W. Carlson, J. McFadden, D.W. Curtis, D. Mitchell, H. Reme, C. Mazelle, J.A. Sauvaud, C. d'Uston, A. Cros, J.L. Medale, S.J. Bauer, P. Cloutier, M. Mayhew, D. Winterhalter, N.F. Ness, Magnetic field and plasma observations at Mars: Initial results of the Mars global surveyor mission, *Science*, **279** (5357): 1676-1680 MAR 13 1998.

Anderson, J. D., G. Colombo, P. B. Esposito, E. L. Lau, and G. B. Trager, The mass, gravity field, and ephemeris of Mercury, *Icarus*, **71**, 337-349, 1987.

Anderson, J. D., R. F. Jurgens, E. L. Lau, M. A. Slade III, and G. Schubert, Shape and orientation of Mercury from radar ranging data, *Icarus,* **124**, 690-697, 1996.

Anselm, A. and G. Scoon, Bepi-Columbo, ESA.s Mercury/Cornerstone Mission, *Planet. Space Sci.,* **49**, 409-420, 2001.

Barabash S, Brandt PC, Norberg O, Lundin R, Roelof EC, Chase CJ, Mauk BH, Koskinen H, Energetic neutral atom imaging by the Astrid microsatellite, *Results of the IASTP Program Advances in Space Research,* **20** (4-5)1055-1060, 1997.

Boyle, C., P. Reiff, M. Hairston, Empirical polar cap potentials, *JGR Space Physics,* **102,** 111-125, 1997.

Bruckner, J. and J. Masarik, Planetary gamm-ray spectroscopy of the surface of Mercury, *Planet Space Sci,* **45**, 39-48, 1997.

Cheng, A.F., R.E. Johnson, S.M. Krimigis, L. Lanzerotti, Magnetosphere, Exosphere, and Surface of Mercury, *Icarus,* **71** (3): 430-440, 1987.

Clark, P., S. McKenna Lawlor, S. Curtis, G. Marr, The Multi-Platform Flyby Approach to Mercury Exploration, *IAC 02,* Q 4.1.04, 2002.

Clark, P. and McFadden, L., New results and implications for lunar crustal iron abundance using sensor data diffusion techniques, *JGR Planets,* 105(E2), 4291-4316, 2000.

Clark, P. and J. Trombka, Remote X-ray spectrometry for NEAR and future missions: Modeling and analyzing Xray production from source to surface, *JGR Planets,* **102**(E7), 16361-384, 1997.

Connerney, J., The magnetic field of Jupiter-A generalized inverse approach, *JGR Space Physics,* **86**, 7679-7693, 1981.

Connerney, J. E. P., and N. F. Ness, Mercury's Magnetic Field and Interior. In *Mercury,* Eds. Vilas, Chapman, and Matthews (Publ. Univ. Arizona Press), 494-513, 1988.

Curtis, S., P.E. Clark, C.Y. Cheung, The Central Role of Reconnection in Space Plasma Phenomena Targeted by the Magnetospheric Multiscale Mission, in *Multiscale Coupling of Sun-Earth Processes,* Edited by A. T. Y. Liu, Y. Kamide, and G. Consolini. (Publ. Elsevier B. V), 2005.

Gampieri, G. and A. Balogh, Mercury's thermoelectric dynamo model revisited, *Planet Space Science,* **50** (7-8): 757-762, 2002.

Glassmeier, K., The Hermean magnetosphere and its ionosphere-magnetosphere coupling, *Planet Space Science,* **45**, 119-125, 1997.

Glassmeier KH, Othmer C, Cramm R, Stellmacher M, Engebretson M, Magnetospheric field line resonances: A comparative planetology approach, *Surveys in Geophysics,* **20** (1): 61-109, 1999.

Gold, R., Solomon, S., McNutt, R., Santo, A., Abshire, J., B., Acuna, M., Afzal, R., Anderson, B., Andreson, G., Andrews, P., Bedini, D.,, Cain, J., Cheng, A., Evans, L., Feldman, W., Follas, R., Gloeckler, G., Goldstein, J., Hawkins, E., Izenberg, N., Jaskulek, S., Ketchum, E., Lankton, M., Lohr, D., Mauk, B., McClintock, W. Murchie, S., Schlemm, C., Smith, D., Starr, R. and Zurbucher, T., The MESSENGER mission to Mercury: scientific payload, *Planet. Space Sci.,* **49**, 467-479, 2001.

Grande, M., Investigation of magnetospheric interactions with the Hermean surface, *Adv. Space Res.* **19** (10) 1609- 1614, 1997.

Grard, R., and A. Balogh, Returns to Mercury: Science and mission objectives, *Planet Space Science*, **49**, 14-15, 1395-1407, 2001.

Grard, R., M. Novara, G. Scoon, BepiColombo-A multidisciplinary mission to a hot planet, *ESA BULL* (**103**), 11-19, 2000.

Hawkins, S., Overview of the multispectral imager on the NEAR spacecraft, *Acta Astro*, **39**, 265-271, 1997.

Hood, L. and G. Schubert, Inhibition of solar wind impingement on Mercury by planetary induction currents, *JGR Space Physics*, **84**, 2641-2647, 1979.

Killen, Rosemary A. and Ip, Wing–H, The surface bounded atmospheres of Mercury and the Moon, *Rev. Geophys.*, **37** (3) 361–406, 1999.

Killen, R.A., M. Sarantos, P. Reiff, Space Weather at Mercury, *Adv Space Research*, **33**, 1899-1904, 2004.

Langevin, Y., Thee regolith of Mercury: present knowledge and implications for a Mercury Orbiter Mission, *Planet Space Sci*, **45**, 31-37, 1997.

McGrath, M.A., Johnson, R.E., and Lanzerotti, L.J., Sputttering of sodium on the planet Mercury, *Nature*, **323,** 694–696, 1986.

Meyer-Vernet and Perche, Tool kit for antennae and thermal noise near the plasma frequency, *JGR.*, **94**, 2405-2415, 1989.

Milkovich, S., J. Head, L. Wilson, Identification of mercurian volcanism: resolution effects and implications for Messenger, *Met Plan Sci*, **37**, 1209-1222, 2002.

Morgan, Thomas H. and Killen, R., A non-stochiometric model of the composition of the atmospheres of Mercury and the Moon, *Planet. Space Sci,.* **45** (1), 81–84, 1997.

Peale, S. J., Does Mercury have a molten core? Nature, 262, 765-766, 1976.

Peale, S. J., Rotational dynamics of Mercury, in *Mercury*, Eds. Vilas, Chapman and Matthews (Publ. Univ Arizona Press), pp. 461-493, 1988.

Petrinec, S. and C. Russell, Intercalibration of solar wind instruments during the international magnetospheric study, *JGR Space Physics*, **98**, 18963-18970, 1993.

Potter, A, C. Anderson, R. Killen, T. Morgan, Ratio of sodium to potassium in the Mercury exosphere, *JGR Planets*, **107**, doi:10.1029/2000JE001493, 2002.

Richardson, J.D., 1987, Ion distributions in the dayside magnetosheaths of Jupiter and Saturn, *JGR Space Physics*, **92**, 6133-6140.

Slavin, J., Mercury's Magnetosphere, *Adv Space Research,* **33**, 1859-1874, 2004.

Solomon, S., McNutt, R., Gold, R., Acuna, M., Baker, D., Boynton, W., Chapman, C., Cheng, A., Gloeckler, G., Head, J., Krimigis, S., M., McClintock, W., Murchie, S., Peale, S., Phillips, R., Robinson, M., Slavin, J., Smith, D., Strom, R., Trombka, J., Zuber, M., The

MESSENGER Mission to Mercury: scientific objectives and implementation, *Planet. Space Sci.*, **49**, 1445-1465, 2001.

Spohn, T., F. Sohl, K. Wieczerkwoski, V. Conzelmann, The interior structure of Mercury: what we know, what we expect from Bepi Colombo, *Planet. Space Sci,* **49**, 1561-1570, 2001.

Sprague, A.L., Kozlowski, R.W.H., and Hunten, D.M., Caloris Basin: An enhanced source for potassium in Mercury/s atmosphere, *Science,* **249**, 1140–1143, 1990.

Strom, R., and G. Neukum, The cratering record on Mercury and the origin of impacting objects, in *Mercury*, F. Vilas, C. Chapman, and, M. Matthews, Eds., U. Arizona Press, 336-373, 1988.

Suess, S. and B. Goldstein, Compression of the hermean magnetosphere by the solar wind, *JGR-Space Physics*, **84**, 3306-3312, 1979.

7.19   SOME QUESTIONS FOR DISCUSSION

1. Discuss the relative merits of identical multi-platform flyby versus the kind of orbiter approach possible now for a mission to Mercury.

2. What advances in technology will the follow-on mission approach recommended here require?

3. Design a lander mission to Mercury, considering environmental requirements and enabling technologies.

Index

Printed in the United States of America